Helmut Acker (Editor)

Oxygen Sensing in Tissues

With 102 Figures and 16 Tables

Springer-Verlag Berlin Heidelberg GmbH

Editor:

Professor Dr. med. Helmut Acker

Max-Planck-Institut für Systemphysiologie
Rheinlanddamm 201, 4600 Dortmund, FRG

ISBN 978-3-642-83446-2   ISBN 978-3-642-83444-8 (eBook)
DOI 10.1007/978-3-642-83444-8

Library of Congress Cataloging-in-Publication Data.

Oxygen sensing in tissues / Helmut Acker, ed.
Bibliography: p. Includes index.

1. Tissue respiration. 2. Oxygen-Metabolism. I. Acker, H. (Helmut), 1939–
QP177.096 1988   574.1'2–dc19   88-12302

© Springer-Verlag Berlin Heidelberg 1988
Ursprünglich erschienen bei Springer-Verlag Berlin Heidelberg 1988

Typesetting: G. Appl, Wemding

Foreword

Since oxygen is mainly transported by diffusion within tissue, the oxygen pressure field reflects the local balance between oxygen supply and oxygen consumption and characterizes the state of oxygen supply. Despite large physiological variations (e.g., hypo- and hyperoxia, hypo- and hypertension, change of energy demand), this oxygen pressure field can remain remarkably constant, demonstrating that very effective mechanisms must exist that guarantee the adequacy of oxygen supply. Today, it is possible to describe in detail the responsible effector mechanisms that produce such a stable state of oxygen supply, but our knowledge of the reactions that sense tissue oxygen supply and trigger the regulatory responses is still incomplete. Since such knowledge is essential for understanding the system of oxygen supply and the way in which it has developed during evolution, even small progress is important.

In this book the important O_2 sensor reactions are discussed as they occur in cells, organs, and organ systems. This broad approach gives an excellent picture of the actual state of knowledge in this field.

Professor Dr. D. W. Lübbers

Contributors

Acker, H.
Max-Planck-Institut für Systemphysiologie, Rheinlanddamm 201,
4600 Dortmund 1, FRG

Bassenge, E.
Institut für Angewandte Physiologie der Universität Freiburg,
Hermann-Herder-Straße 7, 7800 Freiburg, FRG

Bingmann, D.
Poliklinik für zahnärztliche Chirurgie, Universität Mainz, Augustusplatz 2,
6500 Mainz, FRG

Bauer, C.
Physiologisches Institut, Universität Zürich, Winterthurerstraße 190, 8057 Zürich,
Switzerland

Busse, R.
Institut für Angewandte Physiologie, Universität Freiburg,
Hermann-Herder-Straße 7, 7800 Freiburg, FRG

Delpiano, M.A.
Max-Planck-Institut für Systemphysiologie, Rheinlanddamm 201,
4600 Dortmund 1, FRG

Deussen, A.
Physiologisches Institut I, Universität Düsseldorf, Moorenstraße 5,
4000 Düsseldorf 1, FRG

Drews, G.
Institut für Biologie II – Mikrobiologie, Universität Freiburg, Schänzlestraße 1,
7800 Freiburg, FRG

Galle, J.
Institut für Angewandte Physiologie der Universität Freiburg,
Hermann-Herder-Straße 7, 7800 Freiburg, FRG

Gekeler, V.
Physiologisch-Chemisches Institut, Universität Tübingen, Hoppe-Seyler-Straße 4,
7400 Tübingen 1, FRG

Grieshaber, M. K.
Institut für Zoologie, Lehrstuhl für Tierphysiologie, Universität Düsseldorf,
Universitätsstraße 1, 4000 Düsseldorf 1, FRG

de Groot, H.
Institut für Physiologische Chemie I, Universität Düsseldorf, Moorenstraße 5,
4000 Düsseldorf 1, FRG

Grote, J.
Physiologisches Institut, Universität Bonn, Nußallee 11, 5300 Bonn 1, FRG

Heinle, H.
Physiologisches Institut I, Universität Tübingen, Gmelinstraße 5, 7400 Tübingen 1,
FRG

Jelkmann, W.
Institut für Physiologie, Medizinische Universität zu Lübeck,
Ratzeburger Allee 160, 2400 Lübeck, FRG

Knollmann, U.
Institut für Tierphysiologie, Ruhr-Universität Bochum, 4630 Bochum, FRG

Kreutzer, U.
Institut für Zoologie, Lehrstuhl für Tierphysiologie, Universität Düsseldorf,
Universitätsstraße 1, 4000 Düsseldorf 1, FRG

Langer, H.
Institut für Tierphysiologie, Ruhr-Universität Bochum, 4630 Bochum, FRG

Lehmenkühler, A.
Institut für Physiologie, Universität Münster, Robert-Koch-Straße 27 a,
4400 Münster, FRG

Lipinski, H. G.
Neurologische Klinik, Technische Universität, Möhlstraße 28, 8000 München,
FRG

Littauer, A.
Institut für Physiologische Chemie I, Universität Düsseldorf, Moorenstraße 5,
4000 Düsseldorf 1, FRG

Noll, T.
Institut für Physiologische Chemie I, Universität Düsseldorf, Moorenstraße 5,
4000 Düsseldorf 1, FRG

Pietruschka, F.
Max-Planck-Institut für Systemphysiologie, Rheinlanddamm 201,
4600 Dortmund 1, FRG

Pohl, U.
Institut für Angewandte Physiologie, Universität Freiburg,
Hermann-Herder-Straße 7, 7800 Freiburg, FRG

Pörtner, H.O.
Institut für Zoologie, Lehrstuhl für Tierphysiologie, Universität Düsseldorf,
Universitätsstraße 1, 4000 Düsseldorf 1, FRG

Probst, H.
Physiologisch-Chemisches Institut, Universität Tübingen, Hoppe-Seyler-Straße 4,
7400 Tübingen 1, FRG

Rich, I.N.
Abteilung für Transfusionsmedizin, Universität Ulm, DAK-Blutspendezentrale,
Oberer Eselsberg, 7900 Ulm, FRG

Schrader, J.
Physiologisches Institut I, Universität Düsseldorf, Moorenstraße 5,
4000 Düsseldorf 1, FRG

Speckmann, E.-J.
Institut für Physiologie, Universität Münster, Robert-Koch-Straße 27a,
4400 Münster, FRG

Siegel, G.
Institut für Physiologie, Freie Universität Berlin, Arnimallee 22, 1000 Berlin 33,
FRG

Wegener, G.
Institut für Zoologie, Universität Mainz, Kerschensteiner Straße 3, 6500 Mainz 1,
FRG

Weiss, C.
Institut für Physiologie, Medizinische Universität zu Lübeck,
Ratzeburger Allee 160, 2400 Lübeck, FRG

Contents

I. Metabolism

Effect of Oxygen Partial Pressure on Formation of the Bacterial
Photosynthetic Apparatus
(G. Drews)   3

Oxygen Availability, Energy Metabolism, and Metabolic Rate in
Invertebrates and Vertebrates
(G. Wegener)   13

Critical PO_2 of Euryoxic Animals
(M. K. Grieshaber, U. Kreutzer, and H. O. Pörtner)   37

Metabolic and Pathological Aspects of Hypoxia in Liver Cells
(H. de Groot, A. Littauer, and T. Noll)   49

Possible Mechanisms of O_2 Sensing in Different Cell Types
(H. Acker)   65

II. Cell Physiology

Oxygen-Dependent Regulation of DNA Replication of Ehrlich Ascites Cells
In Vitro and In Vivo
(H. Probst and V. Gekeler)   79

Metabolic Events that May Activate Erythropoietin Production in the
Hypoxic Kidney
(C. Bauer)   93

Prostanoids and the Renal Response to Hypoxia
(W. Jelkmann and C. Weiss)   103

Oxygen Tension and Erythropoietin Production: The Role of the
Macrophage in Regulating Erythropoiesis
(I. N. Rich)   113

Effect of Hypoxia on Ca^{2+} Influx and Catecholamine Synthesis in
Chemosensitive Cells of the Carotid Body in Tissue Culture
(F. Pietruschka)   121

III. Heart and Circulation

PO$_2$-Induced Changes of Membrane Potential and Tension in Vascular
Smooth Musculature
(G. Siegel and J. Grote) 131

Possible Function of Endothelial Cells as Oxygen Sensors
(U. Pohl, R. Busse, J. Galle, and E. Bassenge) 143

Influence of Oxidative Stress on Metabolic and Contractile Functions of
Arterial Smooth Muscle
(H. Heinle) 151

Free Cytosolic Adenosine Sensitively Signals Myocardial Hypoxia
(J. Schrader and A. Deussen) 165

IV. Nervous Systems

Changes of the Bioelectrical Activity and Extracellular Micromilieu in the
Central Nervous System During Variations of Local Oxygen Pressure
(E.-J. Speckmann, D. Bingmann, A. Lehmenkühler, and H. G. Lipinski) 179

Possible Meaning of Different Ionic Changes in the Carotid Body During
Hypoxia
(M. A. Delpiano) 193

Oxygen and Glycolysis in the Retina of the Compound Eye of a Crab
(H. Langer, M. Delpiano, U. Knollmann, and H. Acker) 205

Subject Index 213

I. Metabolism

Effect of Oxygen Partial Pressure on Formation of the Bacterial Photosynthetic Apparatus

Gerhart Drews

Institute of Biology 2, Microbiology, Albert-Ludwigs-University, 7800 Freiburg, FRG

Oxygen Is an Important Ecological Factor

Most living chemotrophic organisms generate free energy in the form of ATP by oxidative phosphorylation. This oxygen-consuming process is localized on mitochondrial or bacterial membranes and leads to a membrane potential and a proton gradient over the membrane (proton motive force), which drives ATP synthesis at proton (H^+) ATPase. Oxygen is produced by plant-type photosynthesis during the day when water is split by photosystem II into reducing equivalents for CO_2 fixation and oxygen in a light-energy dependent process.

Molecular oxygen, O_2 (m.p. $-218.4°$ C, b.p. $-182.9°$ C, density 1.429 g/l, m.w. 32.0 g/gmol) is present in the atmosphere at 21% by volume. At normal pressure 760 mm Hg (101.3 kPa) and at 20° C, 0.31 ml (44.29 µg) of oxygen gas is dissolved in 1 ml water. There are many prokaryotic organisms (bacteria and archaebacteria) and very few eukaryotic organisms which can live in the complete absence of oxygen. They may be relics of primitive organisms, living in an early period on the earth when the atmosphere was free of oxygen, or they have been adapted to an anaerobic mode of life. Facultative anaerobic bacteria are able to live under both conditions, aerobically or anaerobically. These bacteria generate ATP in the absence of oxygen by electron transport phosphorylation coupled to anoxigenic photosynthesis. In the presence of oxygen the system of oxidative phosphorylation is active. Some bacteria can respire in the absence of oxygen when a suitable terminal electron acceptor like nitrate or sulfate is present.

Microorganisms, which are exposed regularly to changes of the oxygen partial pressure have developed mechanisms to adapt to the fermentative or respiratory metabolism. When the oxygen partial pressure goes up or down above or below specific threshold values, these organisms are able to sense oxygen and to induce via events of a signal chain a process of cell differentiation which adapts the cell to the respective energy-producing system, so that they can compete with other organisms living in the same niche. Figure 1 is a scheme summarizing results and hypotheses.

Some bacteria are attracted by oxygen. They are chemotactically active. They sense an oxygen gradient and swim towards higher oxygen tensions (Ordal 1985).

In this article the process of membrane differentiation induced by changes of oxygen partial pressure in facultative phototrophic bacteria will be discussed.

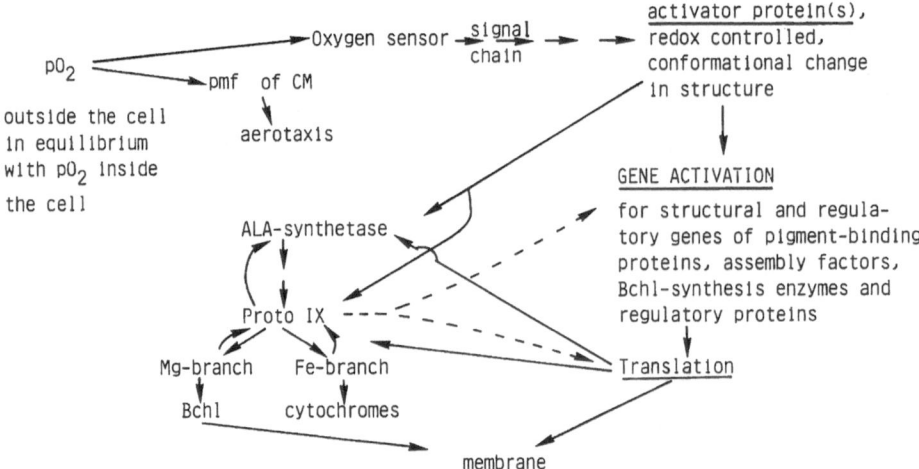

Fig. 1. The effect of oxygen partial pressure (pO₂) on the formation of the pigment proteins of the bacterial photosynthetic apparatus. It is postulated that variations of pO₂ are sensed by an unknown oxygen sensor. The stimulus is transferred via an unknown signal chain to activator proteins which are redox modulated. These proteins activate or inactivate promoters for photosynthetic genes. Possibly gene expression is further regulated by variation of lifetime of mRNA, translation of mRNA, and stable insertion of polypeptides into the membrane. Key enzymes of tetrapyrrol synthesis are regulated on the enzyme level and by regulation of enzyme synthesis. Bchl and precursors affect the expression of genes coding for Bchl-binding proteins and the assembly of the pigment proteins in the membrane. Oxygen tension also regulates aerotaxis. pO₂ directly influences the proton motive force *(pmf)* of the cytoplasmic membrane *(CM)*, pmf determines aerotaxis (swimming towards or opposite a gradient of oxygen tension) *ALA*, aminolevulinate; *Proto IX*, protoporphyrin IX; *Bchl*, bacteriochlorophyll

Bacterial Photosynthesis

In contrast to plants, phototrophic bacteria are anoxygenic. They have no water-splitting and molecular oxygen releasing photosystem II. They have only one photosystem consisting of the photochemical reaction center (RC) and light-harvesting (LH) pigment-protein complexes in addition to the systems for electron and proton transport. The LH or antenna system absorbs photons which create excited states of antenna bacteriochlorophyll (Bchl). These excited singlet states migrate by a random walk over the antennae to the RC. The trapped excitation energy is converted in the RCs into a membrane potential and a redox potential difference. By this power a cyclic electron transport is driven and a proton gradient built across the membrane. Proton gradient and membrane potential are used by different energy-consuming processes, e.g., for production of ATP at the H^+ ATPase or for an active transport across the membrane. Bacterial photosynthesis takes place under anaerobic conditions. (for a review on bacterial photosynthesis, see Dutton 1986).

The Bacterial Photosynthetic Apparatus

In facultative phototrophic bacteria, the photosynthetic apparatus and the respiratory system are not localized in separate cell organelles as in eukaryotic cells, but are bound to the cytoplasmic and intracytoplasmic membrane system (Sprague and Varga 1986; Blankenship and Fuller 1986). Both systems use common electron carriers such as quinones, cytochromes, and membrane-bound enzymes such as cytochrome b/c_1 complex and H^+ ATPase. Specific for the photosynthetic apparatus are the pigment-protein complexes, which are integral macromolecular particles spanning the membrane. The reaction center (RC) consist of three polypeptides having a mol. wt. of about 30000. Two of them, having each five membrane-spanning α helices, bind four molecules of bacteriochlorophyll (Bchl), two molecules of pheophytin, two molecules of ubiquinone, one molecule of carotenoid, and one atom Fe.

The light-harvesting (LH) complexes are oligomers of subunits consisting of two small polypeptides (mol. wt. 5000–8000) binding two or three molecules of Bchl and one molecule of carotenoid. The pigment molecules are noncovalently bound to the proteins. The polypeptides span the membrane by one α helix. The specific and stoichiometric binding of the pigments to the proteins causes an orientation of the axes of the molecules relative to the plan of the membrane and a shift of the long wavelength absorption maxima of Bchl towards the near infrared (800–880 nm). The orientation of the pigment molecules, the optimization of their relative distance to each other, and the overlapping of the near infrared absorption bands make possible an effective energy transfer between the pigment molecules and an efficient energy transduction in the RC. In most bacteria two different LH or antenna complexes are present. Oligomers of the B870 (LHI) complex surround the RC and the LH complex B800–850 (LHII) surrounds in variable amounts and interconnects the RC-LHI complexes (Bachofen and Wiemken 1986; Drews 1985).

Synthesis of the Constituents of the Photosynthetic Apparatus and the Assembly of the Intracytoplasmic Membrane

The photosynthetic apparatus and the intracytoplasmic membrane of facultative photosynthetic bacteria are formed at low oxygen tension in the dark. After lowering of the oxygen tension from about 13 to 0.6 kPa in a chemotrophic dark culture of *Rhodobacter (Rb.) capsulatus* the Bchl content of cells increased about tenfold in 90 min. The pigment-binding polypeptides were synthezised in the same range synchronously and assembled with the pigments in the membrane (Schumacher and Drews 1978). There are specific invagination sites in the cytoplasmic membrane where the components of the photosynthetic apparatus are assembled and the first intracytoplasmic membrane vesicles are formed (Drews and Giesbrecht 1963). At first RC and LHI complexes and later the LHII (B800–850) complexes are formed (Schumacher and Drews 1978). Invagination sites have been isolated as an upper pigmented band in a sucrose density gradient. This fraction is believed to represent early stages of assembly of pigment proteins in the cyto- and intracytoplasmic membrane (Inamine et al. 1984). During a shift experiment from aerobic dark to semiaerobic dark (chemotroph) or to anaerobic light conditions (phototroph) not only RC and LH complexes were synthesized and incorporated into the membrane, but also components of the electron transport system and phospholipids. The incorporation of new components into the membrane system results in an unequal distribution of functional subunits within the cytoplasmic-intracytoplasmic membrane system and determines species-specific intracytoplasmic membrane patterns (Oelze and Drews 1981; Kaufmann et al. 1982; Donohue and Kaplan 1986).

A shift from anaerobic phototrophic to strict aerobic chemotrophic growth conditions results in an abrupt and complete cessation of the synthesis of photosynthetic units and a preferred synthesis of units of the respiratory chain in the cytoplasmic membrane. The celle continue to grow. The area of intracytoplasmic membrane per cell decreases. The process of membrane differentiation is determined by differential rates of incorporation of specific membrane components into the membrane system and not by differential rates of turnover (Oelze and Drews 1981).

Regulation of Membrane Differentiation Under the Influence of Oxygen Partial Pressure

Oxygen tension in the growth medium is the major external factor which determines quantitatively and qualitatively the process of membrane differentiation in facultative phototrophic bacteria. Other factors, as variations of light intensity and substrate composition, effect also this process (Oelze and Drews 1981; Grether-Beck and Oelze 1987).

In the following section the discussion will be restricted to the effect of oxygen tension on the synthesis and assembly of pigment-protein complexes. We should have in mind that the process of membrane differentiation is more complex than described here. Phototrophic bacteria live in a watery surrounding, in lakes, ponds, the sea, or in water-filled soil interstices. The concentration of dissolved molecular oxygen in water depends on temperature, concentration of other solutes, and rates of consumption of oxygen by respiring organisms versus rates of production of oxygen by photosynthetic algae and diffusion of oxygen from air into water. The concentration of oxygen in a lake can change strongly between oversaturation during the day due to oxygen production by photosynthetic active algae in the upper layers and zero when during night or in deep layers of a lake the oxygen is consumed by respiration and a transfer of oxygen from air or oxygen-enriched water layers into the oxygen-exhaustet water body is inhibited by absence of turbulence due to a temperature jump between the upper layers and the lower water body. How do facultative phototrophic bacteria sense oxygen? Movement towards an increasing oxygen gradient (aerotaxis) has been demonstrated in dark cultures of *Rhodobacter (Rb.) sphaeroides*. The velocity of swimming and the tactic response are determined by the proton motive force (electrochemical proton gradient $\Delta\mu H^+$ over the membrane). It is not the electron transport itself, however, that causes the tactic signal (swimming towards higher oxygen concentrations), but instead it is the consequent increase in proton motive force (Armitage et al. 1985). If the electron transport through the respiratory chain is limited by the oxygen concentration, an increase of oxygen tension will increase the proton motive force which triggers a tactic response by preferred swimming toward higher oxygen concentrations. It is unknown whether the sensing of oxygen for aerotaxis is the same system as for a morphogenetic response, i.e., induction of synthesis of pigments and pigment-binding proteins. The old hypothesis that the redox state of a factor (Marrs and Gest 1973), or whether NAD(P)H (Sistrom 1965) or a constituent of the electron transport chain (Cohen-Bazire et al. 1957) regulates key enzymes of the Bchl synthesis pathway, has not been proved rigorously by experimental data. The same is true for the energy charge as a regulatory factor (Gest 1972).

Experiments in many laboratories have shown that the synthesis of pigment-protein complexes is regulated independently from the growth rate. Morphogenesis is induced in cells after lowering of oxygen tension in the exponential or lag phase of growth or in cells in which growth is inhibited by energy or substrate limitation (Drews and Oelze 1981).

The regulation of pigment-protein synthesis by oxygen tension is independent of a functional respiratory chain. Mutants blocked in the respiratory chain are fully inducible in synthesis of pigment-proteins (Marrs and Gest 1973; Hüdig et al. 1986). Changes in rates for respiration and growth do not coincide with changes of rates for Bchl synthesis at specific oxygen partial pressures. Respiratory metabolism is active down to low oxygen tensions of about 100 Pa. Bchl synthesis is initiated when oxygen tension is lower than 650 Pa (*Rhodospirillum (Rs.) rubrum;* Biedermann et al. 1967) or 1.4 kPa (*Rb. capsulatus* but not in *Rs. rubrum* and *Rb. sphaeroides* growing at a relative high oxygen tension of 2.6 to 5.3 kPa; Dierstein and Drews 1974). These data suggest that oxygen regulates Bchl-protein synthesis (Grether-Beck and Oelze 1987).

Recently it has been suggested that the activity of 5-aminolevulinic acid synthetase, the first enzyme of the tetrapyrrol synthesis pathway, is directly regulated by the dithiol-disulfide interchange mediated by the thioredoxin system (Clement-Metral 1986). The thioredoxin system in *Rb. sphaeroides* catalyzes NADPH-dependent reduction of protein disulfides via the following reactions:

1. $\text{Thioredoxin-S}_2 + \text{NADPH} + \text{H}^+ \xrightarrow{\text{TRase}} \text{thioredoxin-S}_2\text{H}_2 + \text{NADP}$
2. $\text{Thioredoxin-S}_2\text{H}_2 + \text{protein-S}_2 \longrightarrow \text{thioredoxin-S}_2 + \text{protein-S}_2\text{H}_2$

The thioredoxin system of *Rb. sphaeroides* has been characterized (Clement-Metral et al. 1986). If this hypothesis is true, we have to ask again which molecule senses O_2 directly? NADP can be reduced by NADH-NADP transhydrogenase and NAD by the respiratory chain or other metabolic processes. The level of reduced nicotinamide adenine dinucleotide in cultures of *Rs. rubrum, Rb. sphaeroides* and *Rb. capsulatus* increased strongly when the oxygen tension dropped below 30–60 Pa (Schön and Drews 1968), while the Bchl synthesis is induced when the oxygen tension increased below 650 Pa (Biedermann et al. 1967). All these biochemical studies support the idea that molecular oxygen does not directly affect the pigment-protein synthesis, but modulates it via the redox-state of sensitive molecules. In order to unravel the complex interplay between oxygen tension and regulation of Bchl-protein synthesis, studies on the gene level as well as biochemical investigations have been initiated.

Regulation of Gene Expression Under the Influence of Oxygen Tension

Lowering oxygen tension below a threshold value strongly induces the formation of messenger (m)RNA coding for pigment-binding proteins of RC and LH-complexes. The level of mRNA for enzymes of Bchl synthesis is increased to a much lesser extent (Biel and Marrs 1983; Klug et al. 1984, 1985; Zhu and Kaplan 1985; Zhu and Hearst 1986). Kinetic studies on mRNA formation, protein synthesis, and assembly of pigment-proteins in the membrane showed clearly that the synthesis of pigment-binding proteins is regulated on the transcriptional level (Klug et al. 1985). The synthesis of the pigments Bchl and carotenoids seems to be regulated mainly on the enzyme level by activation or inactivation of enzymes and only to a small extent on the transcriptional level (Lascelles 1968; Clark et al. 1984; Zhu and Hearst 1986; Viale et al. 1987).

The pigment-binding polypeptides M and L of the RC and α, β of LHI complex (B870) are coded by genes which form the *puf* operon. The *puf* operon is transcribed from more than one promotor. The most active of these promotors is located about 700 bp upstream from the *pufB* gene. This strong promotor is regulated by oxygen tension (Bauer CE, Eleuterio M, Young DA, Marrs BL, unpublished). The investigation of transcription initiation after fusion of specific genes with an "indicator" gene *(lacZ)* has shown that the DNA sequences of the strong promoter, which are located hundreds of nucleotides upstream from the 5' end of the structural genes for *RC-LHI* genes, lose the sensitivity to oxygen regulation when shortened (Beatty et al. 1986). Oxygen regulation of the strong *puf* promoter means activation of transcription initiation when oxygen tension is lowered below a threshold value. The activating DNA-binding protein, which has to be postulated, has not been detected yet.

In *Escherichia coli* the expression of some genes for the anaerobic metabolism seems under the control of the *fnr* gene product (Shaw et al. 1983). Using the *fnr* gene of *E. coli* as a hybridization probe different fragments of the *Rb. capsulatus* genome gave positive signals. A protein of about 30000 mol. wt. was detected in *Rb. capsulatus* cultures with antibodies against the Fnr protein from *E. coli*. This protein was detectable after a shift from anaerobic light to aerobic dark conditions (Waterkamp K, Hüdig H, Drews G, unpublished). The function of this protein remains to be determined.

Addition of levulinic acid after induction of pigment-protein synthesis led to an immediate inhibition of the formation of Bchl and RC and B800–850 specific polypeptides, while the carotenoid synthesis is not impaired (Klug et al. 1986). Levulinic acid, an inhibitor of tetrapyrrol synthesis at an early step, and other inhibitors of Bchl synthesis, did not impair the formation of mRNA specific for RC-870 and B800–850 polypeptides. Mutants of *Rb. capsulatus,* which are blocked at different steps of Bchl synthesis, produce after induction much lower levels of photosynthetic mRNA as the wild-type strain. Bacteria blocked in Bchl synthesis can synthesize pigment-binding polypeptides in very small amounts. These proteins are not stably inserted into the membrane (Klug et al. 1986). These results indicate that Bchl synthesis has a strong influence not only on the stable assembly of Bchl-protein complexes but also on the synthesis of pigment-binding polypeptides. It has been speculated that Bchl precursors of Bchl-associated proteins (Q gene product, Marrs BL, personal communication) effect the expression of genes for pigment-binding proteins (Lascelles 1968; Klug et al. 1986).

The regulation of nitrogenase gene *(nif)* transcription in *Rb. capsulatus* in response to oxygen is determined by the action of DNA gyrase (DNA topoisomerase type II) and DNA topoisomerase type I (Kranz and Haselkorn 1986). The results of experiments with novobiocin suggest that oxygen prevents expression of the positive regulatory gene *nifR4* (Kranz and Haselkorn 1986). The synthesis of the photosynthetic apparatus was also repressed at the same concentration of novobiocin that inhibited *nifH* transcription (Kranz and Haselkorn 1986). It has recently been shown that anaerobically grown cells of *Salmonella typhimurium* have high DNA gyrase activity (high level of DNA supercoiling), whereas aerobically grown cells have high DNA topoisomerase type I activity (high level of more relaxed DNA; Yamamoto and Droffner 1985). In summary, the degree of supercoiling of stretches of DNA, which code for genes expressed under anaerobic conditions, seems to influence the rate of transcription initiation.

Summary and Conclusions

Oxygen partial pressure is the major external factor which in facultative photosynthetic bacteria regulates the formation of both the photosynthetic apparatus and the intracytoplasmic membrane system. The oxygen-sensing system and the events which transfer and convert the signal are unknown. Oxygen seems to regulate the activities of key enzymes for bacteriochlorophyll synthesis by a switch-off mechanism or indirectly by feedback mechanism caused by products of the tetrapyroll synthetic pathway. The synthesis of pigment-binding proteins is regulated on the transcriptional level. Recently it has been shown that at least one promotor for pigment-binding proteins is "oxygen sensitive." The regulatory genes and their products, which are responsible for activation of the structural genes, remain to be detected. It is speculated that redox-sensitive proteins activate or inactivate specific promotors for pigment-binding proteins, for enzymes of bacteriochlorophyll synthesis and assembly factors. Similar to other complex system, such as nitrogenase, the formation of pigment proteins of the bacterial photosynthetic apparatus seems to be regulated by a cascade of processes of gene activation and enzyme regulation.

References

Armitage JP, Ingham C, Evans MCW (1985) Role of proton motive force in phototactic and aerotactic responses of *R. sphaeroides*. J Bacteriol 161: 967–972
Bachofen R, Wiemken V (1986) Topology of the chromatophore membranes of purple bacteria.

In: Pirson A, Zimmermann MH (eds) Photosynthesis III. Encyclopedia of plant physiology. Springer, Berlin Heidelberg New York, pp 620-631

Beatty JT, Adams CW, Cohen SN (1986) Regulation of expression of the rxcA operon of Rhodopseudomas capsulata. In: Youvan DC, Daldal F (eds) Current communications in molecular biology: microbial energy transduction. Cold Spring Harbor Laboratory, pp 27-29

Biedermann M, Drews G, Marx R, Schröder J (1967) Der Einfluß des Sauerstoffpartialdruckes und der Antibiotica Actinomycin und Puromycin auf das Wachstum, die Synthese von Bacteriochlorophyll und die Thylakoidmorphogenese in Dunkelkulturen von Rhodospirillum rubrum. Arch Mikrobiol 56: 133-147

Biel AJ, Marrs BL (1983) Transcriptional regulation of several genes for bacteriochlorophyll biosynthesis in Rhodopseudonomas capsulata in response to oxygen. J Bacteriol 156: 686-694

Blankenship RE, Fuller RC (1986) Membrane topology and photochemistry of the green photosynthetic bacterium Chloroflexus aurantiacus. In: Pirson A, Zimmermann MH (eds) Photosynthesis III. Encyclopedia of plant physiology vol 19. Springer, Berlin Heidelberg New York, pp 390-399

Clark WG, Davidson E, Marrs BL (1984) Variation of levels of mRNA coding for antenna and reaction center polypeptides in Rhodopseudonomas capsulata in response to changes in oxygen concentration. J Bacteriol 157: 945-948

Clement-Metral JD (1986) Regulation of Ala-synthetase by O_2 and thioredoxin system. In: Holmgren A, Bränden C-I, Jörnvall H, Sjöberg B-M (eds) Thioredoxin and glutaredoxin system structure and function. Raven, New York, pp 275-284

Clement-Metral JD, Höög J-O, Holmgren A (1986) Characterization of the thioredoxin system in the facultative phototroph Rhodobacter sphaeroides Y. Eur J Biochem 161: 119-126

Cohen-Bazire G, Sistrom WR, Stanier RY (1957) Kinetic studies of pigment synthesis by non sulfur purple bacteria. J Cell Comp Physiol 49: 25-35

Dierstein R, Drews G (1974) Nitrogen-limited continuous culture of Rhodopseudonomas capsulatus growing photosynthetically or heterotrophically under low oxygen tensions. Arch Microbiol 99: 117-128

Donohue TJ, Kaplan S (1986) Synthesis and assembly of bacterial photosynthetic membrane. In: Pirson A, Zimmermann MH (eds) Photosynthesis III. Encyclopedia of plant physiology, vol 19. Springer, Berlin Heidelberg New York, pp 632-639

Drews G (1985) Structure and functional organization of light-harvesting complexes and photochemical reaction centers in membranes of phototrophic bacteria. Microbiol Rev 49: 59-70

Drews G, Giesbrecht P (1963) Zur Morphogenese der Bakterien-Chromatophoren und zur Synthese des Bacteriochlorophylls bei Rhodopseudomonas sphaeroides und Rhodospirillum rubrum. Zentralbl Bakteriol Parasit Infekt Hyg 190: 508-536

Drews G, Oelze J (1981) Organization and differentiation of membranes of phototrophic bacteria. Adv Microb Physiol 22: 1-92

Dutton PL (1986) Energy transduction in anoxygenic photosynthesis. In: Staehelin LA, Arntzen CJ (eds) Encyclopedia of plant physiology. Photosynthesis III, vol 19. Springer, Heidelberg Berlin New York, pp 197-237

Gest H (1972) Energy conservation and generation of reducing power in bacterial photosynthesis. Adv Microb Physiol 7: 243-282

Grether-Beck S, Oelze J (1987) The development of the photosynthetic apparatus and energy transduction in malate-limited phototrophic cultures of Rhodobacter capsulatus. Arch Microbiol 149: 70-75

Hüdig H, Kaufmann N, Drews G (1986) Respiratory deficient mutants of Rhodopseudomonas capsulata. Arch Microbiol 145: 378-385

Inamine GS, van Houten J, Niederman RA (1984) Intracellular localization of photosynthetic membrane growth initiation sites in Rhodopseudomonas sphaeroidix. J Bacteriol 158: 425-429

Kaufmann N, Reidl H-H, Golecki JR, Garcia AF, Drews G (1982) Differentiation of the membrane system in cells of Rhodopseudomonas capsulata after transition from the chemotrophic to phototrophic growth conditions. Arch Microbiol 131: 313-322

Klug G, Kaufmann N, Drews G (1984) The expression of genes encoding proteins of B800-850 antenna pigment complex and ribosomal RNA of Rhodopseudomonas capsulata. FEBS Lett 177: 61-65

Klug G, Kaufmann N, Drews G (1985) Gene expression of pigment binding proteins of the bacterial photosynthetic apparatus: transcription and assembly in the membrane of *Rhodopseudomonas capsulata*. Proc Natl Acad Sci USA 82: 6485–6489

Klug G, Liebetanz R, Drews G (1986) The influence of bacteriochlorophyll biosynthesis on formation of pigment-binding proteins and assembly of pigment protein complexes in *Rhodopseudomonas capsulata*. Arch Microbiol 146: 284–291

Kranz RG, Haselkorn R (1986) Anaerobic regulation of nitrogen-fixation genes in *Rhodopseudomonas capsulata*. Proc Natl Acad Sci USA 83: 6805–6809

Lascelles J (1968) The bacterial photosynthetic apparatus. Adv Microb Physiol 2: 1–42

Marrs BL, Gest H (1973) Regulation of bacteriochlorophyll synthesis by oxygen in respiratory mutants of *Rhodopseudomonas capsulata*. J Bacteriol 114: 1052–1057

Oelze J, Drews G (1981) Membranes of phototrophic bacteria. In: Ghosh BK (ed) Organization of prokaryotic cell membranes, vol II. CRC, Boca Raton, pp 131–195

Ordal GW (1985) Bacterial chemotaxis: biochemistry of behavior in a single cell. CRC Crit Rev Microbiol 12: 95–130

Schön G, Drews G (1968) Der Redoxzustand des NAD(P) und der Cytochrome b und c2 in Abhängigkeit vom pO2 bei einigen Athiorhodaceae. Arch Mikrobiol 62: 317–326

Schumacher A, Drews G (1978) The formation of bacteriochlorophyll-protein complexes of the photosynthetic apparatur of *Rhodopseudomonas capsulatus* during early stages of development. Biochim Biophys Acta 501: 183–194

Shaw DJ, Rice D, Guest JR (1983) Homology between CAP and *fnr*, a regulator of anaerobic respiration in E. coli. J Mol Biol 166: 241–247

Sistrom WR (1965) Effect of oxygen on growth and the synthesis of bacteriochlorophyll in *Rhodospirillum molischianum*. J Bacteriol 89: 403–408

Sprague SG, Varga AR (1986) Membrane architecture of anoxygenic photosynthetic bacteria. In: Pirson A, Zimmermann MH (eds) Photosynthesis III. Encyclopedia of plant physiology, vol 19. Springer, Berlin Heidelberg New York, pp 603–619

Viale AA, Wider EA, del C Batlle AM (1987) Porphyrin biosynthesis in *Rhodopseudomonas palustris*-XII-aminolevulinate synthetase switch-off/on regulation. Int J Biochem 19: 379–383

Yamamoto N, Droffner ML (1985) Mechanisms determining aerobic or anaerobic growth in the facultative anaerobe *Salmonella typhimurium*. Proc Natl Acad Sci USA 82: 2077–2081

Zhu YS, Kaplan S (1985) Effects of light, oxygen and substrates on steady-state levels of mRNA coding for ribulose-1,5-biophosphate carboxylase and light-harvesting and reaction center polypeptides in *Rhodopseudomonas sphaeroides*. J Bacteriol 162: 925–932

Zhu YS, Hearst JE (1986) Regulation of expression of genes for light-harvesting antenna proteins LH-I and LH-II; reaction center polypeptides RC-L, RC-M, and RC-H; and enzymes of bacteriochlorophyll and carotenoid biosynthesis in *Rhodobacter capsulatus* by light and oxygen. Proc Natl Acad Sci USA 83: 7613–7617

Oxygen Availability, Energy Metabolism, and Metabolic Rate in Invertebrates and Vertebrates

G. Wegener

Institut für Zoologie, Johannes Gutenberg-Universität, 6500 Mainz, FRG

Oxygen and Life on Earth

It has often been emphasized that primitive life originated in an environment devoid of oxygen. The first eukaryotic cells, however, appeared some 1.4 billion years ago when the earth's atmosphere had already turned from a mildly reducing to an oxidizing one by the photosynthetic action of prokaryotes that used H_2O as reducing agent (see Harold 1986, for review). The presence of free oxygen obviously was a major force shaping the evolution of eukaryotic cells. As a consequence all animals are primarily aerobes, using respiratory chains with oxygen as electron acceptor (oxidant) and membrane-bound ATP synthases for the production of ATP.

Even those animals that function in an environment extremely poor in oxygen, such as parasites of the intestinal tract, are dependent on oxygen during certain stages of their life cycles. Moreover, these "champion animal anaerobes" (Hochachka 1980) do not all abandon their mitochondria when living anaerobically, but make use of them in order to increase the yield of ATP, as do other invertebrates that are adapted to temporary lack of oxygen (see below).

Aerobic energy production is so common and widespread because it has substantial advantages compared to anaerobic energy production, including:

1. Aerobic energy production makes use of a variety of substrates, whereas anaerobic metabolism is restricted to the breakdown of carbohydrates, in some instances in combination with certain amino acids.
2. The main foodstuffs, carbohydrates and fat, can be completely oxidized aerobically to harmless and readily excretable end-products, CO_2 and H_2O. In contrast, anaerobic metabolism gives rise to (usually) acidic products and hence protons. The anaerobic cell is burdened with the disposal of these products, which would otherwise interfere with cellular functions.
3. Incomplete anaerobic oxidation does not fully exploit the energy content of substrates, with the consequence that the ATP yield from anaerobic processes is always comparatively meager. Aerobic energy production was a precondition for the evolution of highly active and efficient organisms.

Types of Oxygen Want in Animals and Their Consequences

When oxygen is not available or cannot be used to an extent that ATP hydrolysis is balanced by aerobic ATP synthesis, animals become confronted with several problems. The main problem is a thermodynamic one, the lack of energy to exert physiological functions and to maintain the organism in the state of high order that characterizes life. Another problem is handling the products of anaerobic metabolism.

Hypoxia may affect the whole body or parts of it and it has different effects on different tissues or different animal species. Only very few animals, such as parasitic helminths in the intestinal tract, are adapted to live permanently in an environment very poor in oxygen. In the majority of animals, however, metabolic energy on the whole is produced aerobically. This does not exclude anaerobic energy production in parts of the body or for a limited time span. Even in the presence of oxygen, anaerobic metabolism may permanently be the sole or main source of ATP in particular cells (such as red blood cells which lack mitochondria) or certain tissues (lens and retina of the eye, kidney medulla and some tumors; for review see Krebs (1972); Newsholme and Leech (1983)). Anaerobic products formed in those cells are not excreted but transported to other parts of the body where they are dealt with aerobically.

Temporary anaerobic energy metabolism in aerobic animals can be of two types:

1. It can be brought about by physiological activity, for instance, exercise that requires ATP hydrolysis exceeding aerobic ATP synthesis. This form of anaerobiosis is usually short term and restricted to the working organs. It would either mark a transition phase in which aerobic ATP production is stimulated in the working organ to compensate for the increased ATP turnover or would lead to fatigue ("functional anaerobiosis," because it is a consequence of physiological function; cf. Zebe et al. (1980)). Fatigue would enforce reduction of activity. "Functional anaerobiosis" as such is not a hazard, since regulatory processes (including fatigue) would balance ATP expenditure and ATP production.
2. A potentially dangerous type of anoxia is brought about when aerobic energy production is not sufficient to maintain the basal functions as reflected in the basal (standard) metabolic rate. This form of anaerobiosis often results from limited access to oxygen due to changes in the environment (in this case termed "environmental anaerobiosis"). It could also be due to interference with oxygen transport within the body or to poisoning of aerobic pathways. Because the basal metabolic rate is not met by aerobic ATP synthesis the normal steady state will decay and, if not reestablished by aerobic processes, the organism will succumb.

This review will focus on environmental or related types of anaerobiosis, i.e., temporary anaerobiosis that cannot be avoided by an animal. This type of anaerobiosis offers the greatest threat to an animal and the greatest challenge to research, as very complex responses are elicited at different levels in an organism. The full sequence of events and the mechanisms on which the reactions to oxygen lack are based are not known for any animal, let alone man. Nevertheless, progress has been made in recent years with regard to the understanding of anaerobiosis, by comparing animals that live in different habitats and differ greatly in their ability to cope with temporary lack of oxygen.

Animal Models in the Study of Anaerobiosis

To bring into focus the different aspects of anaerobiosis, various "animal models" will briefly be introduced and some of the results will subsequently be discussed in greater detail.

Mammals are in general not particularly suited as experimental animals for the study of anaerobiosis. Severe hypoxia or anoxia causes dramatic effects in mammals. These effects are due to the extreme sensitivity of the mammalian CNS to anoxia. In humans, for instance, cutting off the brain from its O_2 supply by blocking the bloodflow to the brain (ischemic anoxia) would lead to unconsciousness within 5-7 s and to anoxic seizures some 10 s later (for review see Wegener et al. 1986). Breathing pure N_2 would cause similar effects. This rapid breakdown of coordination of body functions leaves little room for studying the effects of anoxia in intact mammals. Consequently much work has been performed on isolated organs, tissue slices, etc. This approach necessarily has to forego information regarding the body as a whole.

Even mild forms of hypoxia, such as can be produced in decompression chambers, cause profound changes in human reasoning, capabilities, memory, and behavior. During hypoxia a test person would usually not be aware of any mental deficiency and would notice deficiencies only retrospectively when the normoxic condition has been restored (cf. Haldane and Priestley 1935; Ernsting 1965). These subtle effects of hypoxia can hardly be simulated in animal models.

A special case of limited oxygen supply occurs in diving mammals during a dive. As will be shown, a whole array of physiological and metabolic adaptations ("diving response") are elicited by diving in order to avoid anaerobiosis in vital organs, particularly the brain.

Among *lower vertebrates,* interesting experimental animals for the study of anaerobiosis have been found. Lower vertebrates are phylogenetically related to mammals, but have much lower metabolic rates and many species are well adapted to tolerate hypoxia and anoxia. Lower vertebrates are particularly suited to study the effects of anoxia, not only in the various organs but also on the integration of body functions as effected by the nervous and hormonal system.

Among invertebrates, species of the intertidal zone are of special interest as they may regularly encounter hypoxic/anoxic conditions at ebb tide. These animals present themselves for biochemical analyses of their energy metabolism, and metabolic pathways have been investigated extensively and with great success, while integrative aspects and regulatory mechanisms have been studied in less detail.

Parasitic helminths of the intestinal tract and adult *insects* take extreme but opposite positions as to their capacity to function in the absence of oxygen. Many helminths are fully adapted to live permanently in an anoxic environment, while most functions in adult insects are critically dependent on aerobic metabolism.

Effects of Oxygen Want in Vertebrates

Among vertebrates very different capabilities to cope with hypoxia can be found. The effects of anoxia on the common frog will be considered first, then work on other vertebrates will briefly be discussed.

Effects of Anoxia in the Common Frog *(Rana temporaria)*

For the following reasons the common frog appears to be a particularly well-suited experimental animal for the study of the effects of anoxia on body coordination, physiological function, and biochemical and metabolic processes:

1. Frogs have been used for more than 3 centuries in medical and biological research, hence much is known of their biology.
2. Frogs are the right size to allow, with little experimental effort, research both on the whole animal as well as on various organs.
3. Frogs can be handled easily and subjected to anoxia either outside of water or when submerged in water.
4. Frogs have an average tolerance of anoxia and they recover quickly and spontaneously (without artificial reanimation) due to their well-developed skin respiration.

When a frog, rested in a container, is flooded with pure nitrogen instead of air, it remains calm and usually makes no attempt to escape. It then goes through a characteristic sequence of events (cf. Thuy and Wegener 1983; Wegener et al. 1986; Thuy 1987): After a short initial increase (for 2–3 min) the ventilatory rate falls and spontaneous movements are reduced. After about 50 min, spontaneous ventilation ceases and after about 90 min the anoxic frog is not able to maintain its normal body posture. Some 30 min later, none of the various reflexes (cf. Thuy 1987, for details) operate.

When, at this time, frogs are returned to air, recovery is rapid and complete: Ventilatory movements reappear within 2 min and after 5–10 min in air, the animals have assumed their normal posture and all reflexes connected with body posture can be elicited.

Judged from their behavior, frogs appear almost unaffected by 1 h of anoxia. This impression, however, is grossly misleading. Physiological and biochemical processes are impressively changed upon transition to anoxia. The above-mentioned experiment, repeated with a frog rested in a calorimeter, showed a striking decrease in heat production upon transition to anoxic conditions. Heat production decreased by about ⅔ within 30 min and further to about 20% of the control value when anoxia was extended (Wegener et al. 1986; Thuy 1987). Differences in thermodynamic efficiencies (ATP produced/heat dissipated) between aerobic and anaerobic ATP production cannot account for the reduced heat production. Consequently, a marked reduction in energy expenditure (i.e., ATP turnover) must occur in anoxia. Moreover, as heat production of a resting frog reflects the basal (standard) metabolic rate, vital functions must be depressed or suspended during anoxia. Some questions springing from this observation will briefly be discussed in the following paragraphs:

1. Are different organs differently affected by metabolic depression?
2. Which metabolic pathways are operating during anoxia and how are they regulated?
3. Which mechanisms bring about reduction in cellular metabolism?

Heart rate is precipitously decreased upon anoxia, to less than 20% of the control value if anoxia is extended (Fig. 4). The reduced activity of the heart, although

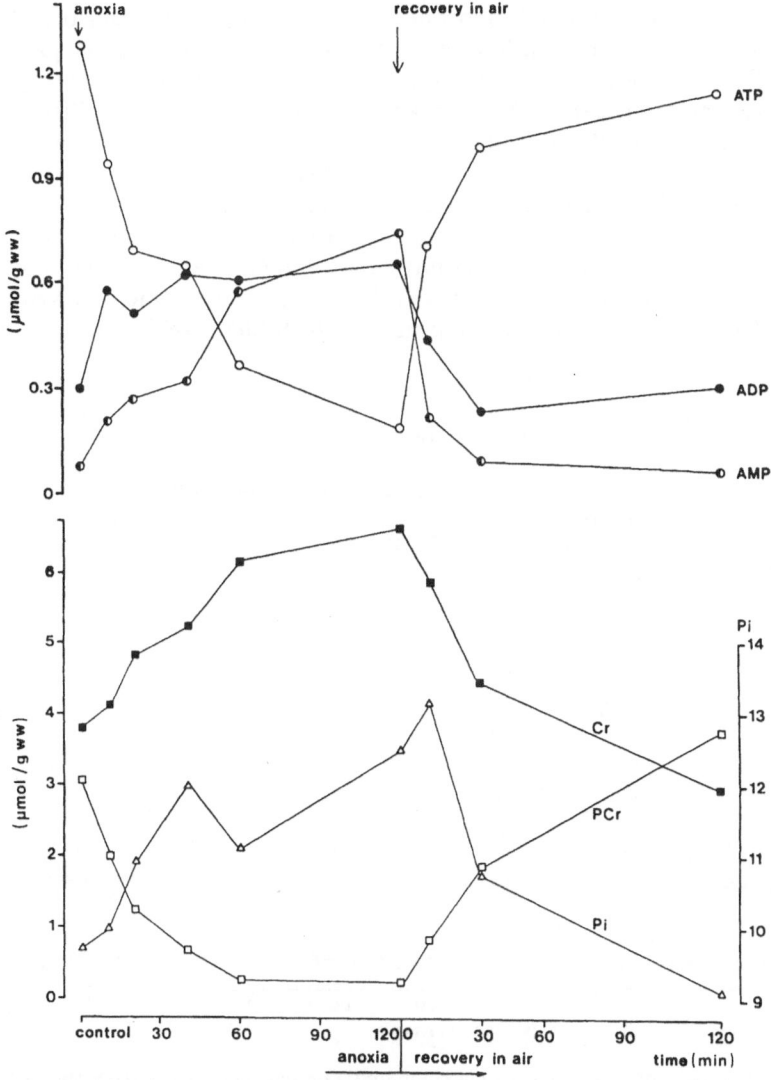

Fig. 1. Effect of anoxia and postanoxic recovery on the contents of phosphocreatine *(PCr)*, creatine *(Cr)*, inorganic phosphate *(Pi)*, and the adenosine phosphates *ATP, ADP,* and *AMP* in brain tissue of the common frog *(Rana temporaria)*. After 120 min of anoxia (at 20°C) the frogs were returned to air for recovery. Each value is a mean from 3-9 individual animals (data from Thuy 1987). For methods see Fig. 6

contributing to the depressed metabolic rate, could only account for a small part
of the metabolic depression. In man, for instance, the heart requires 11% of the total energy expenditure at rest. Reduced heart rate results in reduced perfusion of
the body and, as in other vertebrates, the various organs of frog are probably differently affected. Moreover, in diving mammals or birds, hypoperfusion leads to a
lower temperature and hence to a lower metabolic rate in the hypoperfused parts
of the body. This would, of course, not apply to the ectothermic frog.

Effects of anoxia on various organs have been studied using biochemical
methods. The contents of energy-rich phosphates, some intermediates and products of anaerobic energy metabolism have been measured in various organs, particularly brain (Fig. 1) of the common frog during anoxia and postanoxic recovery
(Thuy 1987; Wegener et al. 1986). In heart and skeletal muscle, no significant
changes in the content of ATP and in the "adenylate energy charge" (AEC) were
detected during 120 min of anoxia, while in liver and kidney the AEC reached a
constant level, after an initial decrease (Fig. 2). In brain, however, ATP and AEC
decreased continuously during the whole anoxic period (Figs. 1 and 2). Consequently, most organs of the frog can temporarily reduce their energy expenditure
during anoxia to an extent that ATP hydrolysis is balanced by anaerobic ATP synthesis. Little is known regarding the mechanisms by which this is achieved.

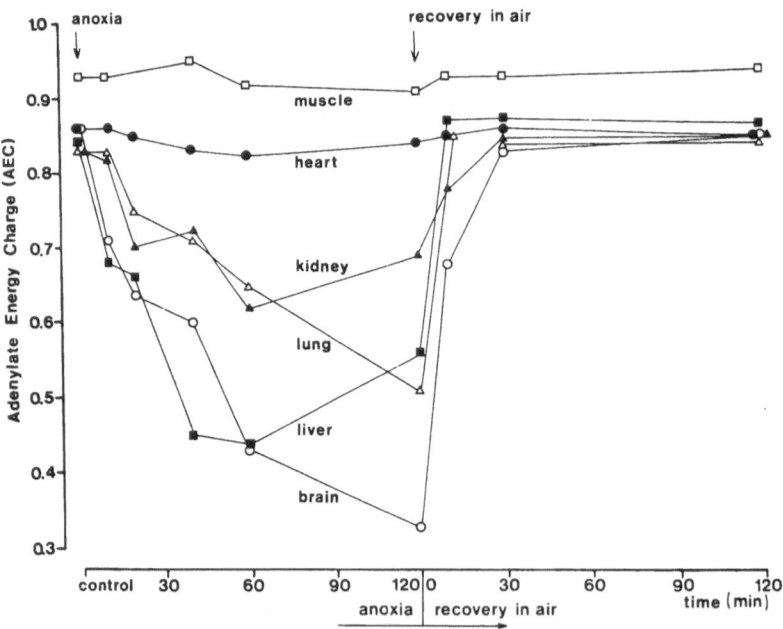

Fig. 2. The effect of anoxia and postanoxic recovery (at 20° C) on the "adenylate energy charge"
(AEC) in organs of the common frog *(Rana temporaria).* AEC is a mean to express the cellular energy state and is defined as $AEC = (ATP + 0.5\ ADP) \cdot (AMP + ADP + ATP)^{-1}$. AEC is hardly affected by anoxia in skeletal muscle and heart; it is decreased to a lower level in kidney and liver.
The continuous decline in AEC in brain during anoxia suggests that the brain is not able to function anaerobically (see text and Fig. 4)

Glycolysis would appear to be the only means of ATP production in the anoxic frog. To maintain function animals would therefore have to increase the glycolytic rate in order to compensate for the less-efficient anaerobic energy production. This inverse relationship of oxygen availability and glycolytic rate is known as the "Pasteur effect." Recent measurements of anoxic heat production and substrate catabolism (for review see Jackson 1968; Pamatmat 1980; Gnaiger 1983; Shick et al. 1983; de Zwaan 1983; Thuy 1987) have shown that animals (vertebrates as well as invertebrates) that are well adapted to tolerate temporary anoxia would rather decrease than increase their glycolytic rate upon anoxia. But different organs react differently and little is known about the regulatory processes that mediate the different responses. In brain, glycogen phosphorylase is activated upon anoxia, in frog brain probably by another mechanism than in other vertebrate brains (cf. Kamp and Wegener 1985), and the glycolytic rate is increased as judged from the increased catabolism of glycogen and the accumulation of lactate (Thuy and Wegener 1983; Thuy 1987). This effect of anoxia has been found in all vertebrate brains studied so far (for references see Wegener et al. 1986).

It is commonly held that the regulatory properties of phosphofructokinase are most important in bringing about the Pasteur effect (Krebs 1972; Siesjö 1978; Sols et al. 1981). AMP and inorganic phosphate, which activate phosphofructokinase, are increased in brain tissue upon anoxia (see Fig. 1). Fructose 2,6-bisphosphate ($F2,6P_2$), however, a recently detected (Hers and van Schaftingen 1982) most-potent activator of animal phosphofructokinases is significantly decreased upon anoxia in brain (and liver) and unaffected in heart (Fig. 3).

In mammalian liver $F2,6P_2$ is crucial in directing glucose metabolism, with high $F2,6P_2$ content being correlated with glucose catabolism via glycolysis, whereas low $F2,6P_2$ levels would indicate glucose production by the liver. In anoxic frog, a decrease in $F2,6P_2$ in liver is accompanied by an increase in blood glucose.

Apart from liver, a physiological function of $F2,6P_2$ has not been established for any vertebrate organ. In this light the changes in the content of $F2,6P_2$ in skeletal muscle during anoxia are of particular interest. There is an initial increase in the content of $F2,6P_2$ in frog muscle. But this increase appears not to be due to anoxia itself, because, when motor nerves were blocked by curare, anoxia did not have an effect on the content of $F2,6P_2$ in skeletal muscle. Since $F2,6P_2$ is a very potent activator of frog muscle phosphofructokinase (Beinhauer, Krause and Wegener, unpublished results) this would suggest that $F2,6P_2$ has an important role in activating phosphofructokinase when muscular contraction is initiated. This hypothesis is in keeping with the rapid and marked increase in $F2,6P_2$ (nearly 50-fold) in leg muscle of intact frogs upon 1 s of swimming, as has recently been observed in the author's laboratory (Kirchgeßner et al. 1987; Krause, Thuy and Wegener, unpublished work).

The changes in the content of $F2,6P_2$ in the anoxic frog would not suggest a major role of $F2,6P_2$ in the regulation of glycolytic rate in frog organs (apart from liver) during anoxia. But in goldfish brain, an increase in $F2,6P_2$ has been found after 24 h of anoxia at 7° C (K. B. Storey, personal communication 1987; see also Storey 1985). Further work has to be awaited before generalizations can be made.

The rapid decrease in ATP content and "adenylate energy charge" (AEC) that

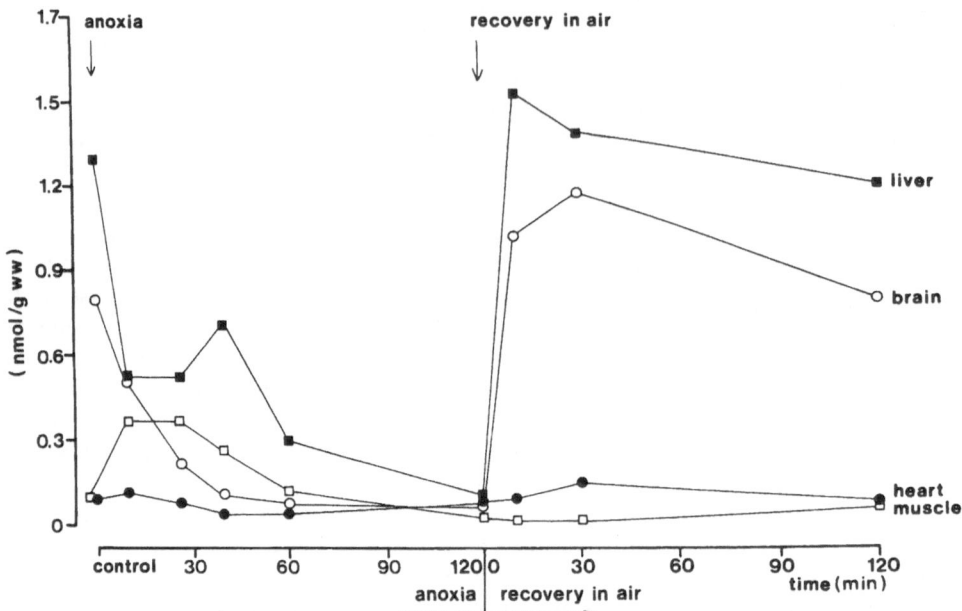

Fig. 3. Effect of anoxia and postanoxic recovery (at 20° C) on the content of fructose 2,6-bisphosphate (F2,6P$_2$) in the organs of the common frog *(Rana temporaria)*. F2,6P$_2$ is reduced upon anoxia in liver and brain and not affected in heart. The initial rise of F2,6P$_2$ in skeletal muscle can be suppressed by curare indicating that it is not caused by anoxia itself, but is likely to reflect nervous activation of the muscle (data from Thuy 1987; Kirchgeßner et al. 1987)

occur in frog brain upon anoxia (see Figs. 1 and 2) has prompted electrophysiological investigation of the function of the CNS during anoxia (Thuy 1987; Wegener et al. 1986). Different parts of the frog CNS are differently affected by anoxia. Cells of the tectum opticum, the optic center in the midbrain, were affected by as little as 10 min of anoxia and ceased to respond to visual stimuli applied to the retina after about 20 min of anoxia (at 20° C) when signals from the optic chiasm could still be collected (see Fig. 4). Thus, the higher centers of frog brain appear most sensitive to anoxia, as has also been demonstrated in the visual system of mammals and man (cf. Ernsting 1965). Spinal cord and spinal nerves appear much more resistent to anoxia; impulse transmission failed after about 90 min of anoxia (Thuy 1987), i.e., at a time when an anoxic frog would lose its normal body posture.

Frogs may encounter severe hypoxia or even anoxia for extended periods in their natural habitat when overwintering at the bottom of ice-covered muddy ponds. Fish such as crucian or goldfish *(Carassius)* can also survive being confined to ice-covered ponds for a long time. Laboratory experiments have confirmed the marked tolerance of anoxia of some fish at low environmental temperatures (for review see Shoubridge and Hochachka 1980, 1983; van den Thillart and van der Waarde 1985). At higher temperatures the anoxic time period that can be survived by crucians is much shorter, less than 12 h at 20° C and only a few hours if the animals are directly transferred to the anoxic medium (Schmidt and Wegener unpub-

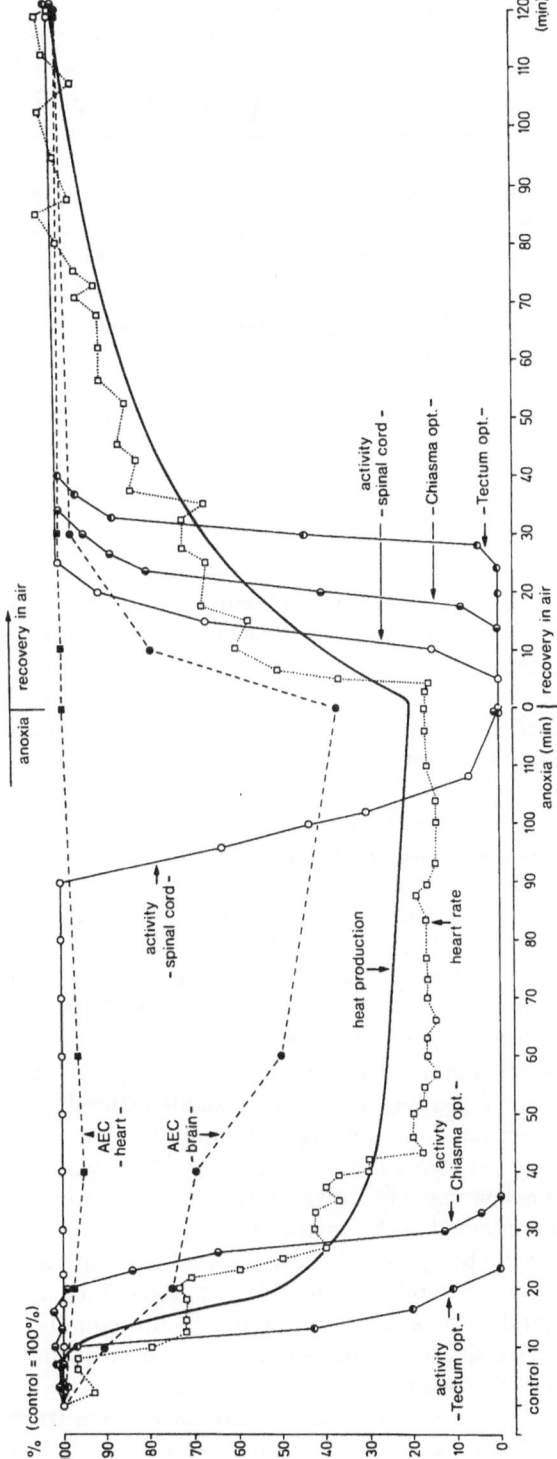

Fig. 4. Effects of anoxia and postanoxic recovery (at 20° C) in brain and heart of the common frog *(Rana temporaria)*. Anoxia leads to a decrease in metabolic rate as reflected in heat production. Heart remains active but at a significantly reduced rate. The function of CNS is lost during anoxia, with the brain being much more sensitive than the spinal cord. The animals recover rapidly from 120 min anoxia (from Thuy 1987)

lished). This clearly indicates that the metabolic rate is a decisive factor in anoxic survival. Any additional stress, such as handling of the animals, that would increase the metabolic rate would decrease the maximum anoxic period that can be survived. Trout, for instance, which in contrast to crucian or goldfish, struggles when exposed to anoxia would rapidly succumb to anoxia (Shepard 1955; Smith and Heath 1980).

Turtles are exceptional in their ability to survive anoxia (for review see Hochachka and Somero 1984; Lutz et al. 1985). For instance, the freshwater turtle, *Pseudemys scripta,* has been reported to withstand anaerobic dives of up to 2 weeks at 16°–18° C. In contrast to frog brain, the content of ATP is rather constant in turtle brain during 120 min of anoxia when the phosphagen stores were already exhausted. This would suggest turtle brain to be able to produce ATP anaerobically in sufficient quantities to balance ATP expenditure during anoxia. ATP turnover is lowered in turtle brain during extended anoxia, but it is not known which functions of the brain are reduced and whether mechanisms are involved that are not available to other vertebrates.

The Diving Response

Mammals and birds are usually not confronted with a shortage of oxygen in their natural habitats. In species adapted to diving, however, oxygen becomes a limiting factor during a dive. The adaptations to diving and the physiological processes during diving have been thoroughly studied in seals (for review see Irving 1939; Scholander 1962; Elsner and Gooden 1983). Seals have a comparatively large blood volume and elevated contents of hemoglobin and myoglobin. As a consequence, seals can store two to three times as much oxygen in relation to body weight as man can (cf. Lenfant 1969), but they can stay submerged more than 10 times as long as man. Bohr, studying diving animals, was puzzled by this phenomenon and noted in 1897 that one is compelled to look for other mechanisms to explain this dicrepancy. It is now known that diving elicits a whole array of mechanisms, collectively called the "diving response."

Bradycardia sets in immediately when a diving animal is submerged (forced diving) and leads to a reduction in heart rate by about 90%, independent of muscular work (e.g., struggling) done by the animal. This form of bradycardia is not brought about by metabolic intermediates but by nervous signals since its onset is immediate and can also be triggered by all sorts of sudden stimuli in seals that are out of water. On the other hand, a submerged animal may not show any bradycardia, provided it is able to emerge for breathing any time it wants.

Although the heart is beating very slowly during a dive, the blood pressure in the central arteries is maintained at the predive level by a potent vasoconstriction that leaves large parts of the body virtually uncirculated. During the dive, the skin and muscle system are ischemic and also blood flow to the kidney and liver practically stops. The peripheral vasoconstriction can also be brought about in seals on land by sudden stimuli like shouting, pinching the skin, or switching the light off. Diving bradycardia and vasoconstriction are executed by vagus nerve and sympathetic nerves, respectively. These can be stimulated via reflexes (receptors of the face, nose, etc.) or metabolic signals (e.g., via the carotid body). Little is known

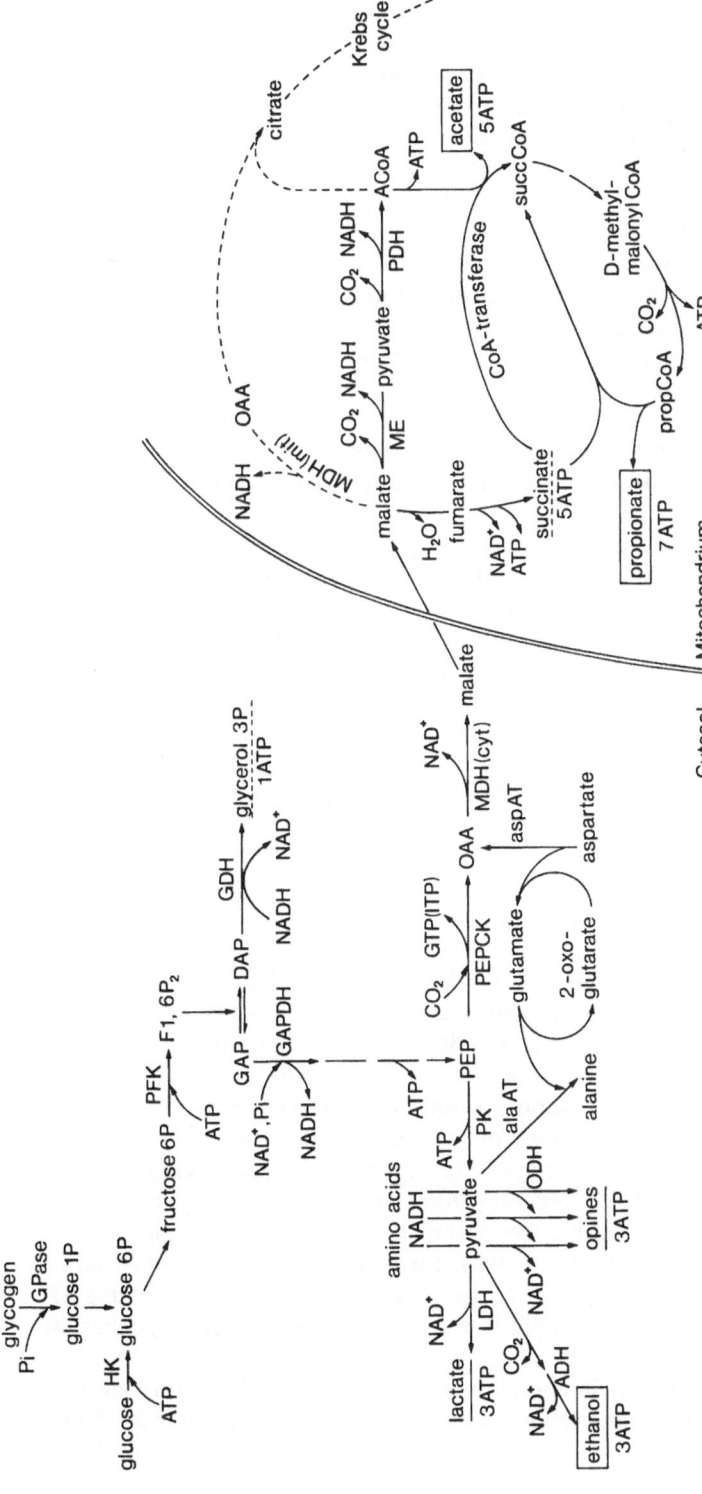

Fig. 5. A simplified general scheme of anaerobic energy metabolism as derived from work on different invertebrates and vertebrates. Not all reactions are working in all species, for instance, in most vertebrates, in echinoderms, and in adult arthropods the anaerobic carbohydrate metabolism appears to be restricted to the classical glycolysis with lactate as the main end-product. In many invertebrate groups, particularly endoparasitic helminths and hypoxia-adapted molluscs and annelids, anaerobic metabolism is extended into the mitochondria, with the consequence that volatile fatty acids can be formed and excreted and additional ATP is gained. Products of anaerobic metabolism are *underlined* (*broken lines* if they can be further metabolized anaerobically) or enclosed in boxes if they can be excreted. The gain of energy-rich phosphates, in terms of ATP, is given with each anaerobic end-product on the simplifying assumption that 1 glucosyl unit derived from glycogen is metabolized anaerobically to yield only the respective end-product. In vivo various end-products are formed simultaneously (for details see text and de Zwaan 1983). *DAP*, dihydroxyacetone phosphate; *GAP*, glyceraldehyde 3-phosphate; *OAA*, oxaloacetate; *PEP*, phosphoenol pyruvate (for further abbreviations see text)

about the higher centers that induce or suppress the diving response (cf. Elsner and Gooden 1983). In ischemic muscle, after O_2 from myoglobin has been depleted, ATP is produced anaerobically from glycogen through glycolysis. The circulating blood is low in lactate because the muscle lactate is only washed out into the general circulation during the postdive recovery when it can be handled aerobically by the liver and heart (and possibly brain).

The excess consumption of O_2 after the dive is much less than would have been expected from the predive resting metabolic rate, often only 20%–25% of the calculated "oxygen dept." Hence, a marked reduction in metabolic rate must occur during asphyxia. This is also reflected in a drop in body temperature (i.e., drop in heat production) during diving.

By means of calorimeters, reduced heat production during asphyxia or anoxia can be measured directly and recent investigations have shown that reduction of metabolic rate is indeed a common response to hypoxia. Elucidation of the mechanisms by which this metabolic depression is brought about is certainly a most challenging problem both with respect to metabolic regulation as well as to possible clinical application.

With all the adaptations to diving, the vital organs, particularly the brain, function aerobically during a dive and the surprisingly dense capillarization of seal brain (Horstmann 1960) may be interpreted in this light. The mammalian brain is not able to function anaerobically, and the diving response may be viewed as a means to avoid this situation. Even those mammals that could stand prolonged apnea will die within minutes when forced to breathe pure nitrogen.

Where it occurs, anaerobic ATP production is apparently restricted to the classical glycolysis in diving mammals. Whether succinate formation (Fig. 5), as has been seen in hypoxic mammalian heart (Wiesner et al. 1986, for review), is of importance in diving remains open.

Effects of Anoxia in Adult Insects

Insects are a highly evolved class of invertebrates, and several of their physiological capacities are unsurpassed in any other animal group. For instance, insects have developed all kinds of locomotion, usually with very high efficiency (for references see Wegener 1988). Most impressive is their ability to fly and this capacity is based on very active energy metabolism and very efficient information processing. Flying insects have the highest metabolic rate of all animals and their working flight muscles have the highest ATP turnover of all muscles. ATP production in flight muscle is completely aerobic. Other organs are also entirely dependent on oxygen; some insect brains, for instance, consume, on a weight basis, more O_2 than any other brain (Wegener 1987, for review). Oxygen is conveyed to the tissues very efficiently by means of air-filled tubes (tracheae), thus taking advantage of the higher content and much faster diffusion of oxygen in air than in aqueous fluids.

In view of their highly aerobic energy metabolism it is hardly surprising that adult insects may react to anoxia even more dramatically than mammals do. Some holometabolous insects, such as the blowfly and honey-bee, become paralyzed

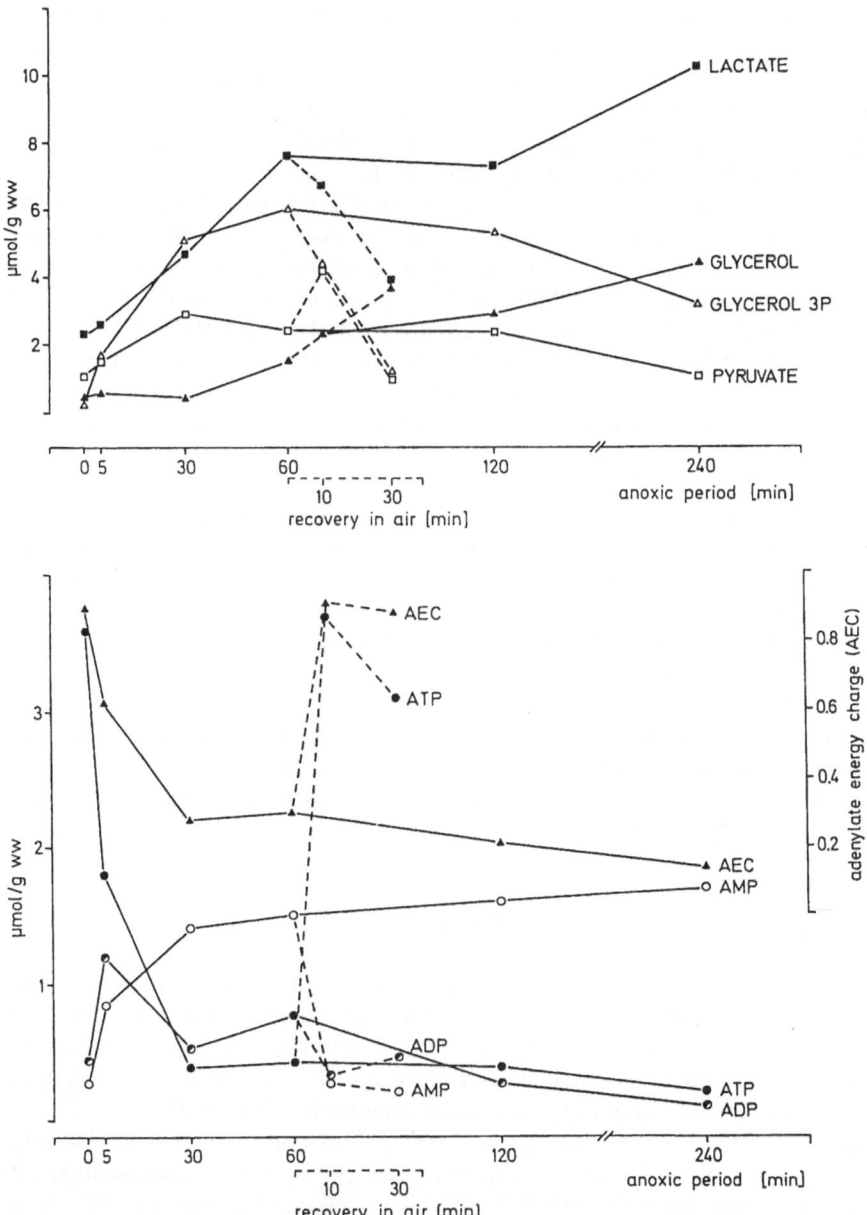

Fig. 6. Metabolic effects of anoxia and postanoxic recovery in the cerebral ganglia of the migratory locust *(Locusta migratoria)*. Female locusts were subjected to pure nitrogen for different intervals at 25° C and then rapidly frozen in liquid nitrogen. The tissue content (in μmol per g wet tissue) of adenosine phosphates and some anaerobic products were measured by using specific enzymatic methods. Anoxia causes a rapid decrease in the adenylate energy charge and a modest increase in anaerobic products. After 1 h of anoxia, the metabolic recovery in brain is rapid. The data presented in this figure are mean values of 3 to 5 determinations (from Wegener et al. 1986)

within seconds when they are flooded with nitrogen. In the locust, the stage of total immotility is reached after a short period of hyperactivity, characterized by escape movements, increased ventilation, loss of body coordination, and anoxic convulsions (see Wegener et al. 1985, 1986; Wegener 1987 for review). Heart beat will rapidly cease and hemolymph flow is at a standstill in anoxic insects.

ATP content and "adenylate energy charge" (AEC) are rapidly decreased upon anoxia in brain tissue of insects as shown in Fig. 6 for the locust. Anaerobic products cannot be removed from the tissue during anoxia as insect organs are not capillarized (open circulatory system) and hemolymph does not circulate. Nevertheless, anaerobic products, of which lactate is predominant in locust, do not rise to high levels during anoxia. In locust brain after 4 h of anoxia, the lactate content is as high as in the ischemic mouse brain after 2 min, and in the anoxic frog brain after 20 min of anoxia, respectively (cf. Wegener et al. 1986). The low activity of lactate dehydrogenase in brain tissue from insects as compared to vertebrates (for review see Wegener 1981, 1987) would certainly contribute to the low glycolytic rate of insect brain during anoxia.

Although the reaction to anoxia is similar in mammals and insects, the capacity of postanoxic revival is not. Most insects revive rapidly from 1 h of anoxia (see Fig. 6) and some species such as housefly or the leaf-cutting ant *Atta sexdens* completely recover from more than 12 h in pure nitrogen at room temperature (see Wegener 1987, for review). Challenging widely held views, these results clearly show that high standard metabolic rate and high neuronal activity in normoxia, as well as rapid loss of neuronal function and a profound decrease in cellular energy state in anoxia do not necessarily lead to rapid irreversible damage from anoxia, as is seen in mammals. Insects are therefore interesting models concerning the question of which metabolic and/or structural properties endow nerve tissue with a high capacity of postanoxic revival.

Anaerobiosis in Marine Invertebrates

Marine invertebrates take up oxygen from water, usually by irrigating internal or external respiratory surfaces which are typically very delicate structures. Most species are not organized to breathe air; when exposed to the atmosphere, respiratory surfaces collapse and the animals become threatened by desiccation. Thus, paradoxically, most marine animals must incur anaerobic ATP synthesis in air, although air contains much more oxygen (20.95%) than water (1% or less). The oxygen content of water is negatively correlated with temperature and salinity. For instance, freshwater of $0°$ C saturated with air at 760 mm Hg contains 10.3 ml O_2 per liter (0.459 mmol, 14.72 mg), and sea water about 20% less (8.0 ml at 35‰ salinity). At $20°$ C, 1 liter of freshwater would hold 6.6 ml O_2 (0.294 mmol), while seawater would only hold 5.4 ml O_2. Consequently, both limited access to aerated water and exposure to air impede respiration and bring about tissue hypoxia or anoxia. Species of low mobility living in the intertidal zone encounter hypoxia regularly at ebb tide. Those animals that live upon the substratum (epifauna) are usually protected from desiccation by tough shells. Shellfish, such as bivalves, gastropods, or barnacles, keep their shells firmly closed in dry air (Newell 1979, for

review). The animals of the infauna (endofauna) tolerate low water while bur-
rowed in the sediment. Because of the peculiarities of their habitat intertidal animals would be expected to be well adapted to temporary hypoxia. Laboratory experiments have indeed demonstrated that these animals can survive complete lack
of oxygen for extended periods.

Few species have been studied thoroughly during the past 15–20 years, among
these the sea mussel *Mytilus,* the lugworm *Arenicola,* and the peanut worm *Sipunculus.* It is now possible to outline the basic properties of their anaerobic energy
metabolism. Moreover, general information about biochemical processes in anoxia has been gathered from these "model animals". Anaerobic metabolism proved
more variable and complex in several invertebrate groups than would have been
expected from the scheme derived from work on vertebrates. However, the differences do not appear fundamental, rather, different animals use different substances and enzymes to reach similar ends. Consequently, the before-mentioned
invertebrates could serve to demonstrate the metabolic problems of anoxia and
their possible solutions. This will be illustrated using the lugworm as an example
and the anaerobic reactions occurring in the lugworm will be integrated into a
general scheme of anaerobic energy metabolism (Fig. 5).

The lugworm *Arenicola marina (Polychaeta)* has been thoroughly studied by
Zebe and collaborators (for review see Zebe et al. 1980; Schöttler et al. 1983; Zebe
and Schöttler 1986; Schöttler 1986). The lugworm is a representative of the marine
endofauna. It lives in L-shaped burrows in muddy sands, a habitat prone to low
oxygen content because of the oxidative metabolism of microorganisms. *Arenicola*
is able to live in deoxygenated deposits when these are immersed by the tide because it pumps, usually at regular intervals, oxygenated water through its burrow.
Oxygen supply of the organs is facilitated by large contractile gills and a closed
circulatory system in wich O_2 is transported by a hemoglobin that has a very high
affinity for O_2, high cooperativity, and a marked Bohr effect. The muscle cells contain myoglobin. Despite these adaptations, lugworms may encounter periods of
oxygen lack quite regularly at ebb tide when aerated seawater is not available.
Laboratory experiments have shown *Arenicola* to recruit anaerobic processes
when the PO_2 in the surrounding water decreases below 45 torr (at 12° C) and to
fall back entirely on anaerobic energy production when the PO_2 is less than 15 torr
(Schöttler et al. 1983).

Experimental anoxia brings about a characteristic sequence of metabolic reactions. During the initial phase of anoxia, *Arenicola* draws upon the phosphagen reserve (which is phosphotaurocyamine) and mobilizes glycogen and the amino acid
aspartate. Anaerobic glycolysis from glycogen yields 3 ATP per glucosyl unit and
gives rise to the production of pyruvate and NADH. Glycolysis can only proceed
if NADH is constantly reoxidized. According to the classical scheme as known
from vertebrates this would be achieved by the activity of lactate dehydrogenase,
and lactate would be the main end product of anaerobic glycolysis. However,
there is very little lactate dehydrogenase in the body wall of *Arenicola,* and lactate
is not accumulated in anoxia. Reoxidation of NADH is achieved by a cytosolic
malate dehydrogenase (see below) and by enzymes that have not been found in
vertebrate tissues. These latter enzymes are called opine dehydrogenases, they catalyze the reductive condensation of pyruvate and an amino acid using NADH as

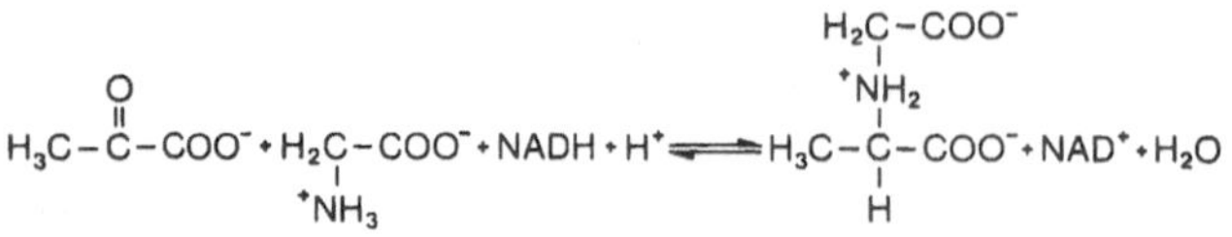

$$H_2C-COO^-$$
$$^+NH_2$$
$$H_3C-\overset{O}{\overset{\|}{C}}-COO^- + H_2\overset{|}{\underset{^+NH_3}{C}}-COO^- + NADH + H^+ \rightleftharpoons H_3C-\overset{|}{\underset{H}{C}}-COO^- + NAD^+ + H_2O$$

Fig. 7.
Reductive condensation of pyruvate and the amino acid glycine to form the opine strombine

H-donor (Fig. 5). In *Arenicola* strombine (Fig. 7) and alonopine are formed from the amino acids glycine and alanine, respectively (Siegmund et al. 1985).

Opine dehydrogenases have high catalytic activities and opines are preferentially formed when the glycolytic rate is relatively high as in the initial phase of environmental anaerobiosis and particularly during exercise (for review see de Zwaan and van den Thillart 1985; Grieshaber and Kreutzer 1986). The body wall musculature of *Arenicola* has a 10-fold higher catalytic capacity to form alanopine than to form strombine, but the tissue content of glycine (about 180 μmol per g) is about 15-fold that of alanine. Formation of alanopine is favored during exercise (digging) when the metabolic rate is high, formation of strombine is favored in environmental anaerobiosis (Siegmund et al. 1985). Opine formation seems to be restricted to some groups of marine invertebrates; the respective enzymes have not been found in arthropods, echinoderms, and vertebrates (for review see Livingstone et al. 1983; Gäde 1983). The biological meaning of opine formation is not fully understood, the reader is referred to recent reviews bearing on this point (de Zwaan and Dando 1984; Grieshaber and Kreutzer 1986).

Aspartate which is also mobilized initially in anoxia is converted to succinate in the following way (see Fig. 5). The NH_2 group of aspartate is transferred to pyruvate (derived from glycogen) via glutamate by the action of aspartate aminotransferase (aspAT, GOT) and alanine aminotransferase (alaAT, GPT; Felbeck 1980), thus yielding alanine and oxaloacetate. Oxaloacetate is reduced to malate by a cytosolic malate dehydrogenase (MDH) and this reaction contributes to the reoxidation of NADH to NAD^+. Malate is not an end-product, but is transported into the mitochondria where it is further metabolized to succinate via fumarate. The formation of succinate from fumarate is dependent on NADH and yields ATP (Schroff and Schöttler 1977). It would appear from experiments using various inhibitors of mitochondrial metabolism that in the formation of succinate the initial parts of the respiratory chain (complex I and II) are involved as indicated in Fig. 8. This would be in accordance with work in mammalian liver mitochondria in which the reverse reaction was demonstrated in that addition of succinate to isolated mitochondria caused the production of NADH and required the hydrolysis of ATP (Klingenberg 1964).

In the initial phase of anaerobiosis, lasting 2–3 h at 12° C, the adenylate energy charge drops from 0.89 to 0.81, although the metabolic rate is rather high (about 40 μmol ATP/h per g dry weight). More than 50% of the ATP turnover can be accounted for by glycolysis, some 20% each would come from phosphagen breakdown and succinate formation (Schöttler 1986).

The situation changes markedly thereafter. The metabolic rate drops precipitously. ATP turnover is reduced to less than 30% between 3 and 12 h and to about

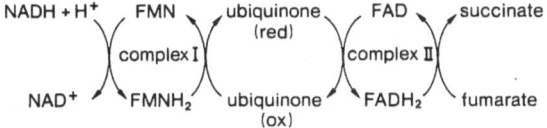

Fig. 8. Proposed flow of hydrogen from NADH in anaerobic mitochondria to yield succinate. Inhibitors of mitochondrial metabolism have indicated that complex I (NADH dehydrogenase) and complex II (succinate dehydrogenase) of the respiratory chain are involved in the formation of succinate. The concomitant synthesis of 1 mol ATP per mol NADH is not a substrate-level phosphorylation, but requires coupled mitochondria and is obviously catalyzed by the F_0F_1ATPase (oxidative phosphorylation), although no O_2 is involved

20% between 12 and 24 h of anoxia, while the energy charge is only slightly decreased during this period (for review see Schöttler 1986).

Parallel to the reduction of the metabolic rate, anaerobic energy metabolism will be restricted to carbohydrate as the only substrate, and glycolysis will be diverted (see below). The most important alteration, however, is recruitment of additional mitochondrial reactions to form excretable anaerobic end-products. In the initial phase of anoxia glycolysis proceeds to form pyruvate from which the main glycolytic products alanine and the opines arise (see Fig. 5). In the later stages of anaerobiosis, glycolysis is diverted at the level of phosphoenolpyruvate (PEP). As before one energy-rich phosphate is gained, but as pyruvate kinase (PK) is not operating no pyruvate appears. Instead oxalacetate is formed from PEP by the enzyme PEP-carboxykinase (PEPCK, see Fig. 5 and Schöttler and Wienhausen 1981). Oxalacetate is reduced to malate, thus regenerating the coenzyme NAD^+, and malate is further metabolized intramitochondrially to give rise to different end-products:

1. Malate can be metabolized to succinate, yielding 1 ATP as has been mentioned before. The NADH necessary for this process can be derived intramitochondrially (see next paragraph and Fig. 5).
2. Part of the malate can be decarboxylated to pyruvate (malic enzyme, ME) and from this can arise acetyl-CoA (pyruvate dehydrogenase, PDH) and finally acetate (at a profit of 1 ATP), which can then be excreted (see Fig. 5). Part of the acetyl-CoA may also be fed into the citric acid cycle.
3. Succinate can be activated to succinyl-CoA by a CoA transferase. Further reactions, including decarboxylation of methylmalonyl-CoA, coupled to the synthesis of ATP, lead to propionyl-CoA which then gives rise to the excretable end-product propionate (Schroff and Zebe 1980), while its CoA moiety can be used to activate another succinate (see Fig. 5).

The effect of anoxia on *Arenicola* has also been studied in vivo using nondestructive and noninvasive methods. Calorimetric measurements have shown a marked decrease in heat production upon anoxia in the lugworm and other intertidal invertebrates (Pamatmat 1980; de Zwaan and van den Thillart 1985, for review). The elegant method of phosphorus nuclear magnetic resonance (^{31}P-NMR) has been used by Kamp (1986) to follow in vivo changes in phosphorus compounds such as ATP, phosphotaurocyamine (PTC), and inorganic phosphate during anoxia and postanoxic recovery (Fig. 9).

G. Wegener

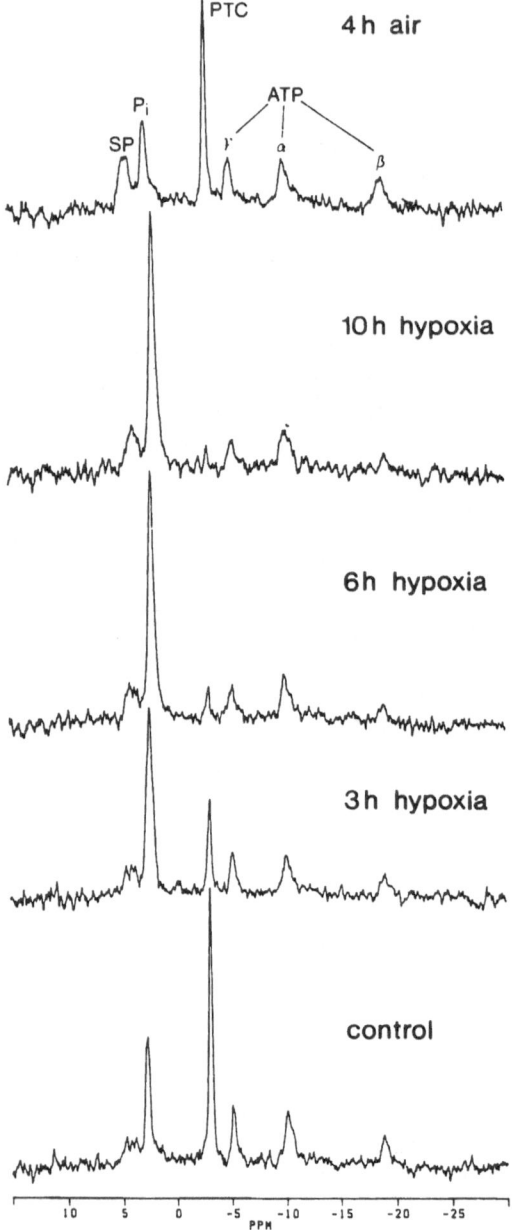

Fig. 9. Effects of hypoxia/anoxia and postanoxic recovery on the contents of phosphagen (phosphotaurocyamine, PTC), ATP, inorganic phosphate (P_i), and sugar phosphates (SP) in the lugworm *Arenicola marina* as studied in vivo using ^{31}P-NMR spectroscopy (from Kamp 1986). A lugworm was kept inside the magnet in a small amount of seawater in a sealed tube at 12 °C, thus producing aggravating hypoxia with time. Spectra, taken after 3, 6, and 10 h of hypoxia/anoxia clearly show the gradual decline in PTC and concomitant increase in P_i. Bubbling air through the tube for 4 h produced the *upper spectrum* (recovery)

Work on anaerobiosis in intertidal animals has identified a complex array of metabolic reactions in anoxia. The capacity for forming various products and operating different metabolic pathways during anaerobiosis is supposed to be of advantage as it allows a flexible response geared to degree and duration of hypoxia. Relatively little, however, is known about how the anaerobic metabolism is regulated. Glycogen, the main anaerobic energy substrate, is obviously mobilized by allosteric activation of glycogen phosphorylase (GPase) mediated by an increase in AMP and inorganic phosphate (see Fig. 9). This is in contrast with "functional anaerobiosis" where covalent modification, i.e., phosphorylation of the inactive b- into the active a-form of the enzyme plays a major role (for review see Kamp 1986).

Another crucial step would be channeling phosphoenol pyruvate into mitochondrial metabolism (see Fig. 5), as occurs in the later stages of anoxia when the glycolytic rate is reduced. This might be due to the high affinity (at a rather low activity) of PEP carboxykinase for its substrates and inhibition of pyruvate kinase by phosphorylation (Holwerda et al. 1983; Plaxton and Storey 1984).

Some observations have en passant been reported suggesting that physiological functions in intertidal animals are affected by anoxia in a specific sequence with the nervous system appearing more sensitive than other organs (cf. Pamatmat 1980). An investigation into this point would help to understand the consequences of anoxia in invertebrates and to elucidate general principles.

Anaerobiosis in Parasitic Helminths

Parasitic helminths of the intestinal tract belong to different animal phyla and they are not related phylogenetically. They are, however, often characterized by complex life cycles, comprising free-living and parasitic stages, and they show interesting similarities in their energy metabolism, which can be regarded as adaptations to the parasitic mode of life (for review see von Brand 1946, 1973, 1979; Saz 1981; Tielens and van den Bergh 1985). These worms bear on our topic because during their life cycles they have to adapt to environments that differ greatly in oxygen availability. It is generally held, although proven only in a few cases, that free-living stages (eggs, miracidia, cercaria) have an aerobic energy metabolism, while the adult stages have a predominantly anaerobic energy metabolism. Parasitic helminths differ in several respects from nonparasitic animals that encounter lack of oxygen. In parasitic helminths the transition from one environment to another is connected with developmental processes, hence the transition is irreversible. As a consequence, the transition from an aerobic to an anaerobic mode of energy production will be permanent. If parasitic worms are confronted with oxygen, they might to some extent (depending on the species) recruit aerobic ATP production. But they would not give up anaerobic metabolism completely.

There are remarkable similarities in energy metabolism between parasitic helminths and nonparasitic animals that encounter temporary anoxia. The differences in detail (such as excretion of lactate or succinate in some parasites) can be regarded as adaptations to a parasitic lifestyle, which is characterized by a relative abundance of foodstuff of high energy content.

All parasitic helminths can perform the classical anaerobic glycolysis to some extent and in some species the predominant end-product is lactate, which is excreted into the medium, i.e., the body fluids or intestinal tract of the host (free-living animals would not excrete lactate). Excretion of ethanol is known of thorny-headed worms *(Acanthocephala)*. In addition to the classical anaerobic glycolysis, most parasitic helminths studied so far have evolved pathways that increase the energetic efficiency of their fermentations (for review see Saz 1981; Tielens and van den Bergh 1985). In these species PEP is carboxylated to oxaloacetate by PEPCK and oxaloacetate is further metabolized to succinate, propionate, and acetate, as has been detailed above and represented in Fig. 5.

Conclusions and Prospects

Apart from some parasites which live permanently in an environment very poor in oxygen, all animals depend on oxygen for survival. Among the various organs, the nervous system, particularly the brain, appears to be most sensitive to hypoxia. Highly active animals with high basal metabolic rates such as higher vertebrates and adult insects encounter a rapid loss of neuronal function in anoxia. Animals that are able to maintain function when oxygen is short in supply apparently do so by making provisions to support the nervous system. One means is sparing energy by suspending functions that are not necessary for immediate survival. This strategy is well illustrated in the "diving response" of birds and mammals. However, a similar response or parts of it can be elicited by hypoxia in all sort of animals, from nondiving mammals and other vertebrates (Thuy and Wegener 1987) to marine molluscs (Feinstein et al. 1977). Therefore, reactions to hypoxia would appear to involve fundamental and phylogenetically old mechanisms, but in most animals next to nothing is known as to how hypoxia is sensed and signalled.

Another intriguing question is how, on the cellular and molecular level, basal functions and energy-consuming reactions are suppressed with the consequence of reducing cellular energy expenditure. Synthesis of macromolecules and other cell constituants is halted in hypoxia and intracellular transport such as axonal transport in neurons is suspended (for review see Siesjö 1978). Some energy-consuming reactions might simply stop because the free energy change ΔG from ATP hydrolysis might not be sufficient. In other reactions more specific mechanisms might be involved (cf. Hochachka and Guppy 1987). Moreover, all anaerobically functioning cells would have to secure anaerobically produced ATP from being split by mitochondrial F_0F_1ATPase which, under normoxic conditions, would primarily function as ATP-synthase. "ATPase inhibitor peptides", small homologous proteins of similar structure that have been found in mitochondria from various sources (ranging from yeast to mammalian organs) might be involved in inhibiting mitochondrial ATPase activity in both environmental and functional anaerobiosis. These ATPase inhibitors bind to the F_1 moiety of the F_0 F_1 ATPase in an energy dependent manner and they are supposed to block its ATPase activity when the electrochemical potential (H^+-gradient) of the inner mitochondrial membrane is low (for review see Schwerzmann and Pedersen 1986). Certainly, the in vivo regulation of mitochondrial F_0 F_1 ATPase in animals of different resistance to anoxia is an important field for future research.

Finally, although a fair picture has evolved of the reactions in anaerobic energy metabolism, little is known about the regulatory mechanisms that direct energy metabolism when oxygen is short in supply.

It is hoped that some of the animal models presented in this paper will provide a further step towards progress in the understanding of the many biological problems connected with oxygen lack.

Acknowledgements. I am much obliged to Dr. Michael Thuy for providing data, to Dipl. Biol. Romi Michel for editorial help and to Mrs. Karin Rehbinder for drawing the figures. Work from the author's laboratory has been supported by the Deutsche Forschungsgemeinschaft (grants We 494/6-3, 7-1 and 9-1)

References

de Zwaan A (1983) Carbohydrate catabolism in bivalves. In: Hochachka PW (ed) The mollusca vol 1. Academic, New York, pp 137-175

de Zwaan A, Dando PR (1984) Phosphoenolpyruvate-pyruvate metabolism in bivalve molluscs. Mol Physiol 5: 285-310

de Zwaan A, van den Thillart G (1985) Low and high power output modes of anaerobic metabolism: invertebrate and vertebrate strategies. In: Gilles R (ed) Circulation, respiration and metabolism. Springer, Berlin Heidelberg New York, pp 166-192

Elsner R, Gooden B (1983) Diving and asphyxia. Cambridge University Press, Cambridge

Ernsting J (1965) The effects of anoxia on the central nervous system. In: Gillie JA (ed) A textbook of aviation physiology. Pergamon, London, pp 270-289

Feinstein R, Pinsker H, Schmale M, Gooden BA (1977) Bradycardial response in *Aplysia* exposed to air. J Comp Physiol 122: 311-324

Felbeck H (1980) Investigations on the role of amino acids in anaerobic metabolism of the lugworm *Arenicola marina*. J Comp Physiol 137: 183-192

Gäde G (1983) Energy metabolism of arthropods and molluscs during environmental and functional anaerobiosis. J Exp Zool 228: 415-429

Gnaiger E (1983) Heat dissipation and energetic efficiency in animal anoxibiosis: economy contra power. J Exp Zool 228: 471-490

Grieshaber MK, Kreutzer U (1986) Opine formation in marine invertebrates. Zool Beitr 30: 205-229

Haldane JS, Priestly JG (1935) Respiration. Clarendon, Oxford

Harold FM (1986) The vital force: a study of bioenergetics. Freeman, New York

Hers H-G, Van Schaftingen E (1982) Fructose 2,6-bisphosphate 2 years after its discovery. Biochem J 206: 1-12

Hochachka PW (1980) Living without oxygen. Harvard University Press, Cambridge

Hochachka PW, Guppy M (1987) Metabolic arrest and the control of biological time. Harvard University Press, Cambridge (Mass.), London

Hochachka PW, Somero GN (1984) Biochemical adaptation. Princeton University Press, Princeton NJ

Holwerda DA, Veenhof PR, Van Heugten HA, Zandee DI (1983) Regulation of mussel pyruvate kinase during anaerobiosis and in temperature acclimation by covalent modification. Mol Physiol 3: 225-234

Horstmann E (1960) Abstand und Durchmesser der Kapillaren im Zentralnervensystem verschiedener Wirbeltierklassen. In: Tower DB, Schadé JP (eds) Structure and function of the cerebral cortex. Elsevier, Amsterdam

Irving L (1939) Respiration in diving mammals. Physiol Rev 19: 112-134

Jackson DC (1968) Metabolic depression and oxygen depletion in the diving turtle. J Appl Physiol 24: 503-509

Kamp G (1986) The mode of glycogen phosphorylase activation during work and hypoxia. Zool Beitr 30: 171-186

Kamp G, Wegener G (1985) Regulatory features of glycogen phosphorylase from frog brain, *Rana temporaria*. J Comp Physiol B 156: 77-85

Kirchgeßner J, Krause U, Thuy M, Wegener G (1987) Der Gehalt an Fructose 2,6-bisphosphat in Organen des Grasfrosches *(Rana temporaria):* Einfluß von Ernährung, Sauerstoff-mangel und Bewegungsaktivität. Verh Dtsch Zool Ges 80: 214-215

Klingenberg M (1964) Reversibility of energy transformations in the respiratory chain. Angew Chem Int Ed Engl 3: 54-61

Krebs HA (1972) The Pasteur effect and the relations between respiration and fermentation. Essays Biochem 8: 1-34

Lenfant C (1969) Physiological properties of blood of marine mammals. In: Andersen HT (ed) The biology of marine mammals. Academic, New York, pp 95-116

Livingstone DR, de Zwaan A, Leopold M, Marteijn E (1983) Studies on the phylogenetic distribution of pyruvate oxidoreductase. Biochem Syst Ecol 11: 415-425

Lutz PL, Rosenthal M, Sick TJ (1985) Living without oxygen: turtle as a model of anaerobic metabolism. Mol Physiol 8: 411-425

Newell RC (1979) Biology of intertidal animals. Marine Ecological Surveys, Fanersham

Newsholme EA, Leech AR (1983) Biochemistry for the medical Sciences. Wiley, Chichester

Pamatmat MM (1980) Facultative anaerobiosis of benthos. In: Tenove KR, Coull BC (ed) Marine benthic dynamics. Belle W Baruch Symp Marine Sci No.11, University South Carolina Press, Columbia, pp 69-90

Plaxton WC, Storey KB (1984) Phosphorylation in vivo of red muscle pyruvate kinase from the channeled whelk, *Busycotypus canaliculatum*, in response to anoxic stress. Eur J Biochem 143: 267-272

Saz HJ (1981) Energy metabolism of parasitic helminths: adaptations to parasitism. Annu Rev Physiol 43: 323-341

Scholander PF (1962) Physiological adaptation to diving in animals and man. Harvey Lect 57: 93-110

Schöttler U (1986) Weitere Untersuchungen zum anaeroben Energiestoffwechsel des Polychaeten *Arenicola marina* L. Zool Beitr 30: 141-152

Schöttler U, Wienhausen G (1981) The importance of the phosphoenol pyruvate carboxykinase in the anaerobic metabolism of two marine polychaetes. In vivo investigations on *Nereis virens* and *Arenicola marina*. Comp Biochem Physiol 68 B: 41-48

Schöttler U, Wienhausen G, Zebe E (1983) The mode of energy production in the lugworm *Arenicola marina* at different oxygen concentrations. J Comp Physiol 149: 547-555

Schroff G, Schöttler U (1977) Anaerobic reduction of fumarate in the body wall musculature of *Arenicola marina* (Polychaeta). J Comp Physiol 116: 325-336

Schroff G, Zebe E (1980) The anaerobic formation of propionic acid in the mitochondria of the lugworm *Arenicola marina*. J Comp Physiol 183: 365-372

Schwerzmann K, Pedersen PL (1986) Regulation of the mitochondrial ATPsynthase/ATPase complex. Arch Biochem Biophys 250: 1-18

Shepard MP (1955) Resistance and tolerance of young speckled trout *(Salvelinus fontinalis)* to oxygen lack, with special reference to low oxygen acclimation. J Fish Res Bd Canada 12: 387-433

Shick JM, de Zwaan A, de Bont AMT (1983) Anoxic metabolic rate in the mussel *Mytilus edulis* L. estimated from simultaneous direct calorimetry and biochemical analysis. Physiol Zool 56: 56-63

Shoubridge EA, Hochachka PW (1980) Ethanol: novel end product of vertebrate anaerobic metabolism. Science 209: 308-309

Shoubridge EA, Hochachka PW (1983) The integration and control of metabolism in the anoxic goldfish. Mol Physiol 4: 165-195

Siegmund B, Grieshaber MK, Reitze M, Zebe E (1985) Alanopine and strombine are end products of anaerobic glycolysis in the lugworm *Arenicola marina* (Annelida, Polychaeta). Comp Biochem Physiol 82B: 337-345

Siesjö BK (1978) Brain energy metabolism. Wiley, Chichester

Smith MJ, Heath AG (1980) Response to acute anoxia and prolonged hypoxia by rainbow trout

(Salmo gairdneri) and mirror carp *(Cyprinus carpio)* red and white muscle: use of conventional and modified metabolic pathways. Comp Biochem Physiol 66 B: 267–272

Sols A, Castaño JG, Aragón JJ, Domenech C, Lazo PA, Nieto A (1981) Multimodulation of phosphofructokinases in metabolic regulation. In Holzer H (ed) Metabolic interconversion of enzymes. Springer, Berlin Heidelberg New York, pp 111–123

Storey BK (1985) A re-evaluation of the Pasteur effect: new mechanisms in anaerobic metabolism. Mol Physiol 8: 439–461

Thuy M (1987) Wirkung von Sauerstoffmangel auf Stoffwechsel und Organfunktionen bei Wirbeltieren. Eine vergleichend biochemisch-physiologische Untersuchung unter besonderer Berücksichtigung des Grasfrosches *Rana temporaria*. Dissertation, Johannes Gutenberg-Universität, Mainz

Thuy M, Wegener G (1983) Anoxieeffekte im Energiestoffwechsel des Grasfrosches. Hoppe Seyler's Z Physiol Chem 364: 1224–1225

Thuy M, Wegener G (1987) Effekte von Hypoxie auf die Aktivität und den Energiestoffwechsel des Herzens bei Wirbeltieren. Verh Dtsch Zool Ges 80: 229

Tielens AGM, Van den Bergh SG (1985) The (an)aerobic energy metabolism of parasitic helminths. Mol Physiol 8: 359–369

Van den Thillart G, Van der Waarde A (1985) Teleosts in hypoxia: aspects of anaerobic metabolism. Mol Physiol 8: 393–409

von Brand T (1946) Anaerobiosis in invertebrates. Biodynamica, Normandy 21, Missouri

von Brand T (1973) Biochemistry of parasites. Academic, New York

von Brand T (1979) Biochemistry and physiology of endoparasites. Elsevier, Amsterdam

Wegener G (1981) Comparative aspects of energy metabolism in nonmammalian brains under normoxic and hypoxic conditions. In: Stefanovich V (ed) Animal models and hypoxia. Pergamon, Oxford, pp 87–109

Wegener G (1987) Insect brain metabolism under normoxic and hypoxic conditions. In: Gupta AP (ed) Arthropod brain: its evolution, development, structure and functions. Wiley, New York, pp 369–397

Wegener G (1988) Elite invertebrate athletes: flight in insects, its metabolic requirements and regulation and its effect on life-span. (to be published)

Wegener G, Michel R, Kieffer S, Thuy M (1985) Different metabolic reactions to anoxia in the central nervous system of lower vertebrates and adult insects and their possible significance for postanoxic revival. Mol Physiol 8: 653–655

Wegener G, Michel R, Thuy M (1986) Anoxia in lower vertebrates and insects: effects on brain and other organs. Zool Beitr 30: 103–124

Wiesner RJ, Rüegg JC, Grieshaber MK (1986) The anaerobic heart: succinate formation and mechanical performance of cat papillary muscle. Exp Biol 45: 55–64

Zebe E, Schöttler U (1986) Vergleichende Untersuchungen zur umweltbedingten Anaerobiose. Zool Beitr 30: 125–140

Zebe E, Grieshaber M, Schöttler U (1980) Biotopbedingte und funktionsbedingte Anaerobiose. Biol Unserer Zeit 10: 175–182

Critical PO$_2$ of Euryoxic Animals

M. K. Grieshaber, U. Kreutzer, and H. O. Pörtner

Institut für Zoologie, Lehrstuhl für Tierphysiologie, Universität Düsseldorf, 4000 Düsseldorf 1, FRG

Critical PO$_2$

During normoxic conditions most animals have enough oxygen available to maintain an aerobic energy metabolism. This holds especially true in animals that live in air, where a low oxygen supply is hardly ever encountered except for at high altitudes (Bouverot 1985). Animals inhabiting aquatic habitats may be more often subjected to moderately or severely hypoxic conditions, since water contains less oxygen and several ecological factors such as stratification, salinity, temperature, or plant respiration may further reduce oxygen availability. Therefore, it is not surprising that aquatic species in particular have acquired several physiological and biochemical adaptations to cope with environmental hypoxia.

Many experiments have been carried out in which the effects of a decreasing ambient PO$_2$ on several parameters such as the rate of oxygen consumption, blood flow, or ventilatory rate have been monitored. In an attempt to generalize the deluge of data obtained, Fry (1957) distinguished fish species with an oxygen consumption independent of a declining ambient PO$_2$ from those with an oxygen uptake which changes in relation to ambient oxygen tension. Prosser (1973) extended this point of view and tried to classify animals into oxyregulators (i.e., O$_2$-independent) and oxyconformers (O$_2$-dependent). As may be expected from such an oversimplified classification, several investigators have questioned this system (e.g., Mangum and van Winkle 1973). Many exceptions have been noted and attributed to, e.g., pigment affinity, level of activity, temperature, or even the means of measurements.

Oxyregulating animals decrease their oxygen consumption rapidly below a certain species-specific range of ambient PO$_2$. This PO$_2$ is traditionally called the critical oxygen tension (P$_c$). The physiological significance of the critical PO$_2$ has been a matter of controversy for a long time. Some authors have assumed that the P$_c$ marks the point at which the respiratory pigment is not fully saturated when it leaves the gills (Walshe 1950; Redmond 1955). Others, like Hughes (1973) have speculated that at the P$_c$ the respiratory pump is no longer able to pass sufficient oxygen over the gills to meet the aerobic demands of an animal. The latter assumption, however, implies that at the P$_c$ at least some tissues must become anaerobic, a view which was already proposed by R. E. Young in 1973. Whichever explanation of the P_c is accepted, it is obvious that it cannot be fixed at a certain range of oxygen tension even within the same species, because as the demand of the tissue for oxygen increases, e.g., due to enhanced activity, the minimum ambient PO_2 at which adequate oxygen is obtained must be higher.

Since oxygen consumption increases during enhanced activity, one could speculate that the value of the P_c is also affected. In addition is has been widely accepted that under these conditions the heavily working muscle, e.g., the white skeletal muscle of vertebrates, and some fast-twitch muscles of invertebrates become hypoxic. Some recent experiments prove, however, that one has to distinguish whether the muscle tissue itself becomes hypoxic or whether the metabolic flux through the Embden-Meyerhof pathway increases to such an extent that the mitochondria cannot consume all reducing equivalents and pyruvate for final oxidation. Reoxidation of NADH must then occur in the cytosolic compartment and, therefore, one might expect an anaerobic energy metabolism in the cytoplasm with the mitochondria still working aerobically.

In order to prove whether the P_c marks the transition from aerobiosis to anaerobiosis, first of all a detailed understanding of the anaerobic metabolism during ambient lack of oxygen as well as during enhanced activity is required. Later on the concept of the P_c is reevaluated and its applicability extended from oxyregulating to oxyconforming animals.

Anaerobic Metabolism

Anaerobiosis of course is best evoked in an animal if it is exposed to extreme hypoxia or anoxia. The experimental set up is simple although precautions have to be taken to completely eliminate oxygen. Anaerobiosis is then best reflected by the onset of anaerobic energy production, which can be assessed by monitoring several key metabolites (Gäde and Grieshaber 1988) since the tissue content of some compounds drastically changes upon the onset of anoxia. During the past 15 years several groups of investigators have dealt with this topic publishing many facets of anaerobiosis. Several reviews summarize these data and give a comprehensive view of anaerobic energy metabolism (De Zwaan 1983; Hochachka and Somero 1984; Kreutzer et al. 1985; Gäde and Grieshaber 1986).

Since some knowledge of the biochemical reactions is necessary for the understanding of the physiological meaning of P_c, a brief account of the essentials of an anaerobic energy metabolism should be presented here. This process is described in the same sequential order as it is thought to occur in hypoxic animals: (a) the utilization of a phosphagen, (b) the fermentation of glycogen as well as some amino acids, and (c) the transformation of succinate to volatile fatty acids within mitochondria if anoxia prevails. It should, however, be mentioned that volatile fatty acids are only produced in some facultatively anaerobic invertebrates and some parasites (Kluytmans et al. 1975; Saz 1981; Pörtner et al. 1984).

As soon as oxygen is not supplied sufficiently at the mitochondrial level every animal makes use of its phosphagen to buffer the content of ATP. Phosphagens are well suited for an instant supply of ATP, since they are directly connected by the reaction:

$$\text{phosphagen} + \text{ADP} + \text{H}^+ \rightleftharpoons \text{aphosphagen} + \text{ATP}$$

A wide variety of phosphagens which are all guanidine derivatives can be extracted from animal tissues (Van Thoai and Roche 1964). In vertebrates only creatine phosphate can be found. In insects, crustaceans, and molluscs phosphoarginine occurs, whereas in some annelids phospotaurocyamine or phospholombricine are present. It seems that those muscles contain a high level of phosphagen which are able to perform sudden and extreme work.

Enzymes catalyzing the transphosphorylation of a phosphagen (phosphagen kinases) usually show high activities and it is likely that phosphagen kinases catalyze close to their reaction equilibrium (Beis and Newsholme 1975). A slight increase in the ADP content will, therefore, immediately initiate ATP synthesis. It is for these reasons that phosphagens can serve as a primary energy source providing a substantial amount of ATP during the initial phase of anaerobiosis. Due to the limited amount of phosphagen available, however, it is usually rapidly exhausted.

Glycogen and some amino acids can also serve as substrates for anaerobic energy metabolism. Glycogen is a major energy source, since it can be stored in large amounts within the cell. The glycogen content in mammalian muscle cells can account for approximately 60 to 120 μmol glycosyl units$\cdot$g^{-1} wet wt (Hultman 1967; Kepler and Decker 1969). In some marine invertebrates average gylcogen contents of about 100 μmol glycosyl units$\cdot$g^{-1} wet wt have been reported (De Zwaan and Zandee 1972). Glycolysis via the Embden-Meyerhof-Parnas pathway leads to the formation of ATP and pyruvate which in the absence of oxygen must be reduced to balance the cytosolic redox ratio.

Most animals thus far investigated possess at least one NADH-dependent pyruvate reductase activity. In vertebrates this is lactate dehydrogenase. In a variety of marine invertebrates different pyruvate reductase activities are known which reduce pyruvate in the presence of an amino acid and NADH giving rise to an opine (Gäde and Grieshaber 1986). Specifically the following reactions can occur:

1. Pyruvate + NADH $\rightleftharpoons$ D,L-lactate + NAD$^+$
2. Pyruvate + glycine + NADH $\rightleftharpoons$ strombine + NAD$^+$ + H$_2$O
3. Pyruvate + L-alanine + NADH $\rightleftharpoons$ alanopine + NAD$^+$ + H$_2$O
4. Pyruvate + L-arginine + NADH $\rightleftharpoons$ octopine + NAD$^+$ + H$_2$O
5. Pyruvate + taurine + NADH $\rightleftharpoons$ tauropine + NAD$^+$ + H$_2$O

Lactate and the four opines are considered to be end-products of anaerobic glycolysis which are accumulated in the cytoplasm during the early phase of long-term anaerobiosis, as well as during enhanced activity (Grieshaber and Kreutzer 1986).

Some amino acids can serve as another fuel for the anaerobic energy production (Fig. 1). Generally, several amino acids show particularly high concentrations in marine invertebrates. During the early phase of environmental anaerobiosis a decrease in the aspartate content and concomitant increase in the alanine level has been established. Aspartate is metabolized in the cytosol via the following reactions:

1. Aspartate + 2-oxoglutarate $\rightleftharpoons$ glutamate + oxaloacetate
2. Glutamate + pyruvate $\rightleftharpoons$ alanine + 2-oxoglutarate
3. Oxaloacetate + NADH $\rightleftharpoons$ malate + NAD$^+$

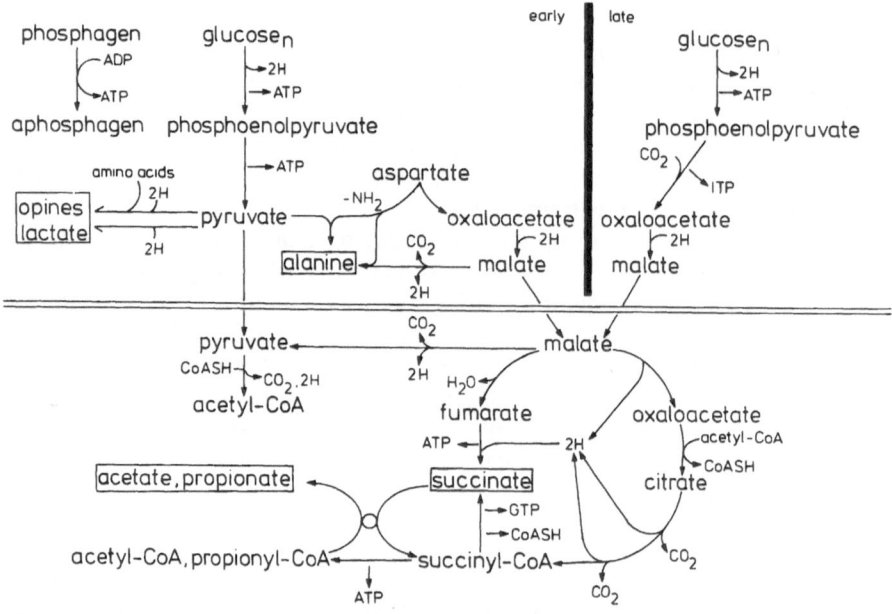

Fig. 1. Scheme of anaerobic energy metabolism as found in facultatively anaerobic marine invertebrates starting from aspartate and glycogen in the early phase and solely from glycogen in the late phase of anaerobiosis. End-products and some intermediary products (outlined in *boxes*) are formed in the cytosol and inside the mitochondria. The mechanism indicated for the release of acetate and propionate from the CoA esters is based on Schöttler (1986)

This reaction sequence which is catalyzed by glutamate oxaloacetate transaminase (1), glutamate pyruvate transaminase (2), and malate dehydrogenase (3) integrates aspartate and glycogen fermentation resulting in alanine and malate formation (Fig. 1).

The cytosolic end-products of the early phase of anaerobiosis are lactate, opines, alanine, and malate. In most animals, however, the malate content only slightly increases compared with other end-products. Malate, in contrast, serves as the metabolite which links the anaerobic energy metabolism of cytoplasm and mitochondria not only during the early phase of anaerobiosis but also during prolonged deprivation of oxygen.

In the good anaerobes such as many annelids and molluscs the degradation of glycogen to lactate and opines ceases after 3–5 h because the major route of the metabolic flux does not lead to pyruvate. Instead, it deviates from the gylcolytic pathway at the phosphoenolpyruvate branchpoint where phosphoenolpyruvate is now carboxylated to oxaloacetate with the concomitant production of a nucleotide triphosphate. Redox balance in the cytosol is achieved by the reduction of oxaloacetate to malate.

The control of the relative flux through the carboxylation of phosphoenolpyruvate to oxaloacetate versus the transphosphorylation of phosphoenolpyruvate to pyruvate is exerted via the regulation of the activity of pyruvate kinase. Holwerda et al. (1981) as well as Holwerda et al. (1983) demonstrated the existence of two interconvertible forms of pyruvate kinase in the adductor muscle of *Mytilus edulis*. They found that with progressive hypoxia one variant of the pyruvate kinase became predominant. It is characterized by a lower enzymatic activity, decreased affinity for phosphoenolpyruvate, and an increased sensitivity to inhibition by L-alanine and protons. In *Busycotypus canaliculatum* the changes in the catalytic properties of pyruvate kinase could be assigned to the phosphorylation of a threonine residue in the protein. This may lead to conformational changes resulting in kinetic differences between normoxic and hypoxic forms of pyruvate kinase. As a consequence, phosphoenolpyruvate could be metabolized preferentially to oxaloacetate (Plaxton and Storey 1984).

Malate arising from the metabolization of either aspartate or glycogen enters the mitochondria (Fig. 1). If the intramitochondrial redox potential is highly negative as can be expected during oxygen deprivation, malate is dehydrated to fumarate which is further converted to succinate with the concomitant rephosphorylation of one mole ADP per mole of succinate formed. This reaction sequence which represents a reversal of some intermediary steps of the normoxic citric acid cycle immediately leads to an increase of the succinate concentration as soon as oxygen is limiting. An elevated succinate level will also persist during prevailing lack of oxygen. Thus, succinate is a true indicator of anaerobically working mitochondria.

Particular interest has focussed on the reduction of fumarate to succinate since the aerobic and anaerobic function of the respiratory chain meet in this reaction. During normoxia reducing equivalents are transferred from succinate via FAD to oxygen, whereas during anoxia electrons are transferred from NADH via site 1 of the respiratory chain to fumarate. A stoichiometric synthesis of approximately one mole of ATP per mole of reduced fumarate has been verified by several investigators (Schöttler 1977; Schroff and Schöttler 1977). Unfortunately, determinations of the lowest oxygen partial pressure at which a reversal of the succinate dehydrogenase reaction commences have not yet been obtained. It can, however, be expected that the P$_{50}$ of fumarate reduction is in the order of magnitude of the P$_{50}$ of cytochrome oxidase, i.e., that fumarate reduction can only occur if the electron transfer to oxygen is completely inhibited. This corroborates again the view that an increase in succinate content should provide the most sensitive tool for monitoring mitochondrial hypoxia and many experiments have indeed proven that the tissue content of succinate increases as soon as oxygen is limiting (Table 1).

For a long time it was also assumed that an increase in lactate concentrations within the tissues (in particular muscle) or blood signals lack of oxygen. This view was derived from the fact that hypoxia which was experimentally imposed on an animal provokes lactate accumulation. Experiments with exercising vertebrate muscles, however, have indicated lactate accumulation while cytochrome aa$_3$ was still oxygenated (Jöbsis 1963). Aerobic lactate formation not only occurs in white muscle but also in submaximally exercising red muscles of vertebrates where mitochondria are abundant (Connett et al. 1984).

With succinate as an easily measurable indicator of mitochondrial anoxia at hand, it should be possible to distinguish between an anaerobic metabolism in the cytosol (as it may occur either during enhanced muscular activity or during ambient lack of oxygen) versus mitochondria synthesizing ATP anaerobically. For an

Table 1. Levels of succinate found during anaerobiosis in invertebrate and vertebrate tissues (in μmol g^{-1} fresh weight, $\bar{x} \pm$ SD)

Species	Tissue	Succinate content		
		Control	Anaerobiosis	(h)
Sipunculus nudus[a]	Body wall musculature	0.10 ± 0.01	1.46 ± 0.08	(24)
Mytilus edulis[b]	Post. add. muscle	0.07	3.04	(24)
Arenicola marina[c]	Body wall musculature	0.05 ± 0.10	2.60 ± 0.20	(24)
Scrobicularia plana[d]	Whole animal	0.14 ± 0.03	3.17 ± 0.27	(24)
Felis[e]	Papillary muscle in vitro	0.13 ± 0.08	0.77 ± 0.52	(1)

Data are from [a] Pörtner et al. (1984); [b] DeZwaan et al. (1982); [c] Schöttler et al. (1984); [d] Brinkhoff et al. (1983); and [e] Wiesner et al. (1986)

invertebrate nonstriated muscle a segregation of an anaerobically working cytoplasmic energy metabolism from mitochondrial respiration could be proven. The isolated introvert retractor muscle of the peanut worm *Sipunculus nudus* was stimulated to contract isometrically under normoxic and under hypoxic conditions (Kreutzer et al. 1985). Under either experimental procedure energy is derived from the transphosphorylation of phospho-L-arginine as well as from anaerobic gylcolysis terminating in the formation of octopine. The kinetics of phosphagen degradation and octopine formation are almost indentical under normoxia and hypoxia, respectively (Fig. 2). At first sight this could be attributed to a poorly developed system for oxygen transport causing intracellular hypoxia even in normoxic muscles. Figure 3, however, shows that succinate production as the indicator of mitochondrial anoxia is barely detectable in normoxic muscles, while it is prominent in muscles working in a nitrogen atmosphere. Thus, the stimulation of anaerobic ATP production in the cytoplasm appears to be independent of mitochondrial oxygenation. In the described example (a white muscle) aerobic energy production might not be limited by the mechanism of oxygen transport, but rather by the overall capacity of the mitochondria to regenerate ATP.

Succinate, however, is only a transient end-product of anaerobiosis. If anoxic conditions prevail the new steady-state concentrations of succinate are maintained. Some of it is released from the tissue into the body fluid. Moreover, in the good anaerobes among invertebrates succinate is further metabolized to propionate (Fig. 1). As soon as the succinate pool is sufficiently high, succinate is transformed to succinyl-CoA. In the early phase of mitochondrial anoxia acetyl-CoA serves as donor of the CoA moiety, but when the propionyl-CoA content in the mitochondria increases, the latter compound activates succinate to succinyl-CoA. This reaction, which is catalyzed by a CoA transferase, does not require any additional energy. Succinyl-CoA is carboxylated via L-methylmalonyl-CoA to D-methylmalonyl-CoA which in turn is decarboxylated by the propionyl-CoA carboxylase to propionyl-CoA. The reaction also renders one mole ATP per mole of propionyl-CoA formed. This metabolite again swaps the CoA moiety with succinate leaving propionate as the final end-product of anaerobiosis. The latter is released into the medium, and therefore, does not disturb the internal milieu of the

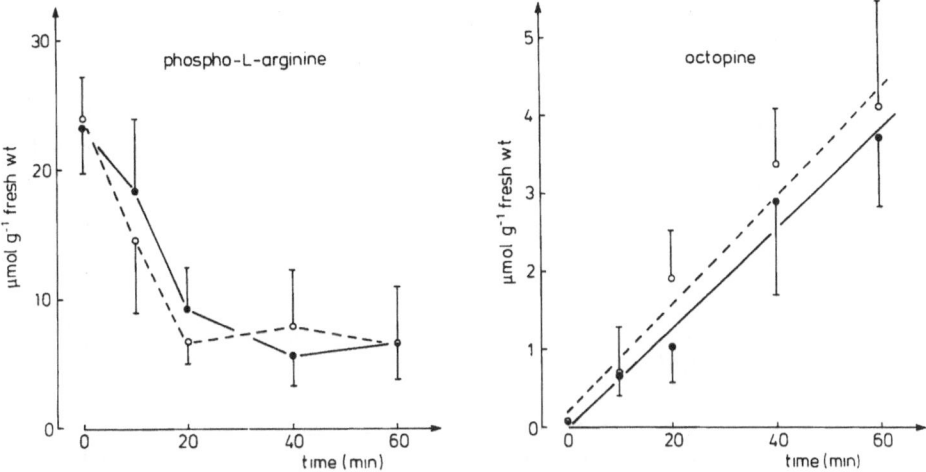

Fig. 2. Phospho-L-arginine and octopine content (μmol$\cdot$g^{-1} fresh weight; $\bar{x}\pm$SD, $n=4$) of isolated introvert retractor muscle (IRM) of *Sipunculus nudus* contracting isometrically at a frequency of 6 min^{-1} (modified after Kreutzer et al. 1985). ●——● contraction under normoxic conditions (PO$_2=150$ torr), O-----O contraction under hypoxic conditions (PO$_2<10$ torr)

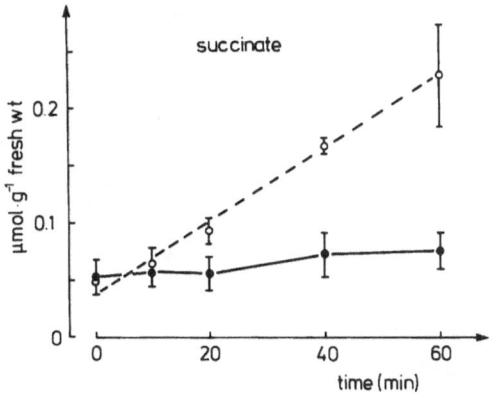

Fig. 3. Succinate content of isolated IRM contracting isometrically under normoxia and hypoxia (legend see Fig. 2)

cell even during prolonged anoxic conditions (Schultz and Kluytmans 1983; Schöttler 1986).

Using this knowledge of metabolic reactions occurring during anaerobiosis, we can determine at which ambient PO$_2$ an animal becomes anaerobic simply by using the cytosolic end-products of anaerobic glycolysis and succinate as probes.

Both mitochondrial and cytosolic anaerobiosis were investigated in the oxyconformer *Sipunculus nudus* (Pörtner et al. 1985). Specimens of *Sipunculus nudus* were incubated for 24 h at various ambient PO$_2$, followed by the determination of anaerobic end-products in extracts of the body wall musculature. As shown in Fig. 4, there is a significant increase in opine and succinate levels at a PO$_2$ of 20 Torr. The onset of end-product accumulation occurs below the same PO$_2$ value

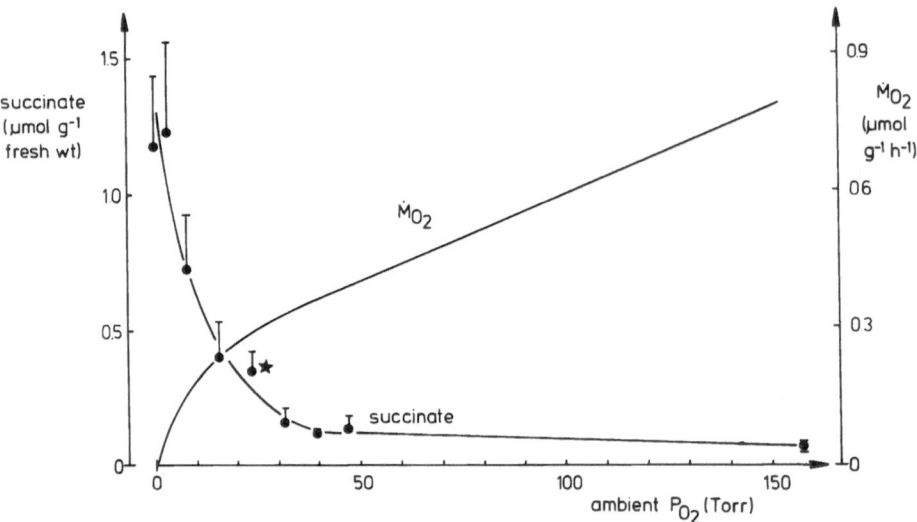

Fig. 4. Succinate contents in the introvert retractor muscles of *Sipunculus nudus* after 24 h of exposure to different ambient oxygen tensions ($\bar{x} \pm$ SD; modified after Pörtner et al. 1985) compared with the oxygen consumption curve. Under progressive hypoxia the introvert retractor muscles as an "inner" tissue are first affected by a reduction in oxygen supply. The onset of anaerobiosis in these tissues is reflected by a concomitant drop in the rate of oxygen consumption decline for the whole animal

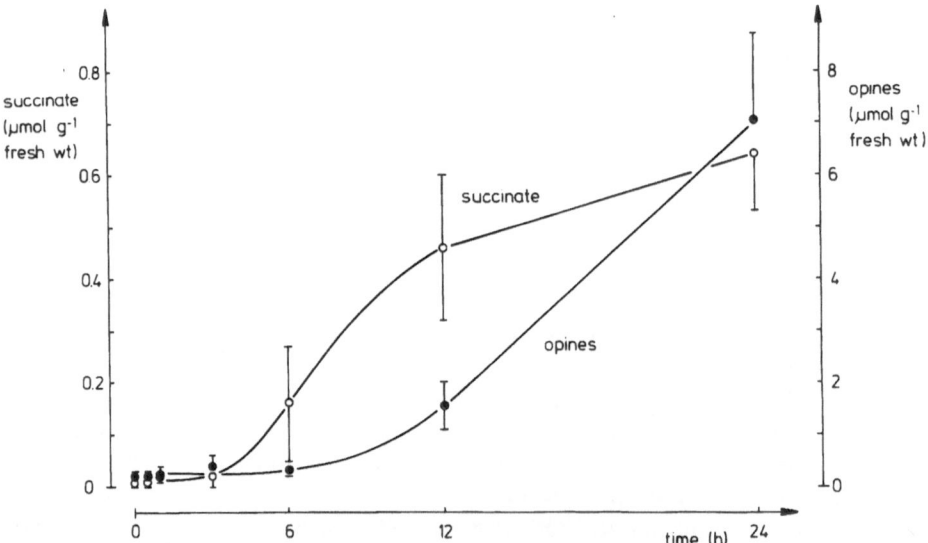

Fig. 5. Succinate and opine (sum of octopine, strombine, alanopine) contents in the body wall musculature of *Sipunculus nudus* after different periods of hypoxic exposure at $PO_2 = 7.5$ torr ($\bar{x} \pm$ SD; modified after Pörtner et al. 1985). The significant onset of succinate accumulation occurs before the formation of opines becomes apparent

below which the oxygen consumption of the worm decreases rapidly. These data already indicate that the P$_c$ can not only be defined for an oxyregulator, but also for an oxyconforming animal (see below).

After having determined the PO$_2$ below which anaerobiosis commences, an even lower PO$_2$ was chosen to evaluate whether the cytosolic end-products or the mitochondrial succinate levels increase first. Since almost quiescent animals were used muscular work was minimal. Under these circumstances the mitochondria should be able to accept all the pyruvate resulting from the Embden-Meyerhof-Parnas pathway as long as oxygen is available. When the mitochondria gradually become deprived of oxygen, then succinate levels should increase. Since not all of the pyruvate can be metabolized by the mitochondria anymore, an increase of opine levels should follow, depending on the equilibrium of the respective opine dehydrogenase reaction. Figure 5 demonstrates that the content of succinate increases already after 6 h, while opine levels are increased after 12 h. The results of this experiment lead to the assumption that below the P$_c$ the onset of cytosolic and mitochondrial metabolism is interdependent, whereas during muscular activity no such interrelated control is found.

Critical Oxygen Tension(s), Metabolic Rate, and Anaerobiosis

The variety of responses to declining oxygen tensions described in the literature leads us to search for a concept which allows the explanation of similarities and differences in these observations. A general parameter which would allow such a comparison is the lowest possible rate of aerobic metabolism which has to be defined for both oxyregulators and oxyconformers. For oxyregulators the lowest rate of oxygen consumption at a given PO$_2$ is defined as the standard metabolic rate (Beamish 1964; Ultsch et al. 1980). Since several oxyregulators (e. g., among crabs and fish) have been shown to maintain this rate down to PO$_2$ values below which anaerobiosis starts (Teal and Carey 1967; Pamatmat 1978; Pelster et al. 1988), SMR would represent the metabolic rate closest to the lowest aerobic metabolic rate.

SMR is evaluated in long-term measurements (Ultsch et al. 1980) or by extrapolation to zero activity (Beamish 1964). Ideally, the measurement in a flow-through system is required in order to make sure that the lowest rate of metabolism at a given PO$_2$ is considered. For extrapolation to zero activity, anaerobic metabolism must be excluded as a source for energy during the different degrees of activity. Since only the influence of locomotory activity is considered by the extrapolation procedure, the calculated rate of oxygen consumption still includes the influence of ventilation. The methodological problems involved in the evaluation of SMR may be one reason why this analysis has not yet been done systematically for many species throughout the animal kingdom. Considering the variety of experimental procedures applied, it may very well be that some animals presently seen as oxyconformers will then be identified as oxyregulators with a low standard metabolic rate.

In oxyconformers, the lowest possible rate of aerobic metabolism cannot easily be derived from oxygen consumption curves. Considering, however, the same methodological concerns as described for oxyregulators, the PO_2 can be evaluated, below which anaerobic metabolism becomes involved in energy production (Pamatmat 1978; Schöttler et al. 1983; Pörtner et al. 1985). The rate of oxygen consumption found at this PO_2 can be seen as the lowest aerobic metabolic rate or, as the standard metabolic rate (SMR) of an oxyconforming animal. Below this critical PO_2 a more rapid drop in oxygen consumption is very likely to occur. This change in the rate of oxygen consumption decline could be demonstrated to coincide with the onset of anaerobic metabolism in the oxyconformer *Sipunculus nudus* (Pörtner et al. 1985; Fig. 3). The rapid fall of the rate of oxygen consumption reminds of the pattern of oxygen uptake variation seen below the P_c in oxyregulators.

In *Arenicola marina* two critical PO_2 values can be distinguished, Pc_I which indicates the shift towards anaerobic metabolism (Schöttler et al. 1983) and a higher Pc_{II} which is characterized by the transition from oxyconformity towards oxyregulation (Toulmond and Tchernigovtzeff 1984).

It is diffusion limitation which very likely leads to the onset of anaerobiosis below the Pc. The predominant view is that diffusion limitation of oxygen in the cytoplasm together with the rate of mitochondrial respiration and the clustering of mitochondria in areas of high energy needs define the point at which the oxygen concentration finally becomes limiting for the rate of oxygen uptake of a cell (Jones 1986). This PO_2 may vary between cells and tissues. It depends on the rate of perfusion of a tissue and also on the density of capillaries and on SMR. For the whole organism, it may depend on the structure of the circulatory system, on the O_2 affinity of the pigment and its regulation, and on the function and structure of the gas exchange system.

Respiratory control of mitochondria in a wide range of ambient PO_2 above the Pc very likely does not depend on diffusion limitations for O_2, since they only require a PO_2 of approximately 0.05 torr to saturate their oxidative capacity (Chance 1976). O_2 measurements at the site of O_2 consumption are required to substantiate this hypothesis. In a more detailed approach it has to be analysed, whether the critical PO_2 of some mitochondria is reached at the same PO_2 in vivo as the critical PO_2 of the whole animal.

Acknowledgements. We thank the Deutsche Forschungsgemeinschaft and Fonds der Chemischen Industrie for support (Gr 456/9-4 and Po 278/1-1)

References

Beis I, Newsholme EA (1975) The contents of adenine nucleotides, phosphagens and some glycolytic intermediates in resting muscles from vertebrates and invertebrates. Biochem J 152: 23–32

Beamish FWH (1964) Respiration of fishes with special emphasis on standard oxygen consumption. III. Influence of oxygen. Can J Zool 42: 355–366

Bouverot P (1985) Adaptation to altitude-hypoxia in vertebrates. In: Farner DS (ed) Zoophysiology, vol 16. Springer, Berlin Heidelberg New York

Brinkhoff W, Stöckmann K, Grieshaber MK (1983) Natural occurrence of anaerobiosis in molluscs from marine habitats. Oecologia 57: 151–154

Chance B (1976) Pyridine nucleotide as an indicator of the oxygen requirements for energy-linked functions of mitochondria. Circ Res [Suppl 1] 38: 31–38

Connett RJ, Gayeski TEJ, Honig CR (1984) Lactate accumulation in fully aerobic, working dog gracilis muscle. Am J Physiol 246: H120–H128

De Zwaan A (1983) Carbohydrate catabolism in bivalves. In: Hochachka PW (ed) The Mollusca, vol 1. Metabolic biochemistry and molecular biomechanics. Academic, New York, pp 137–175

De Zwaan A, Zandee DI (1972) Body distribution and seasonal changes in the glycogen content of the common sea mussel *Mytilus edulis*. Comp Biochem Physiol 43A: 53–58

De Zwaan A, De Bont AMT, Verhoven A (1982) Anaerobic energy metabolism in isolated adductor muscle of the sea mussel *Mytilus edulis*. J Comp Physiol 149: 137–143

Fry FEH (1957) The aquatic respiration in fish. In: Brown ME (ed) Physiology of fishes, vol 1. Academic, New York, pp 1–63

Gäde G, Grieshaber MK (1986) Pyruvate reductases catalyze the formation of lactate and opines in anaerobic invertebrates. Comp Biochem Physiol 82B: 255–272

Gäde G, Grieshaber MK (1988) Measurement of anaerobic metabolites. In: Bridges CR, Butler PJ (eds) Techniques in comparative respiratory physiology – an experimental approach. Cambridge University Press, Cambridge

Grieshaber MK, Kreutzer U (1986) Opine formation in marine invertebrates. Zool Beitr NF 30: 205–229

Hochachka PW, Somero GN (1984) Biochemical adaptation. Princeton University Press, Princeton, NJ

Holwerda DA, Kruitwagen ECD, De Bont AMT (1981) Regulation of pyruvate kinase and phosphoenolpyruvate carboxykinase activity during anaerobiosis in *Mytilus edulis* L. Molec Physiol 1: 165–171

Holwerda DA, Veenhof PR, Van Heugten HAA, Zandee DI (1983) Modification of mussel pyruvate kinase during anaerobiosis and after temperature acclimation. Molec Physiol 3: 225–234

Hughes GM (1973) Respiratory responses in fish. Am Zool 13: 475–489

Hultmann E (1967) Studies on muscle metabolism of glycogen and active phosphate in man with special reference to exercise and diet. Scand J Clin Lab Invest [Suppl 94] 19: 1–63

Jöbsis FF (1963) Spectrophotometric studies on intact muscle. II. Recovery from contractile activity. J Gen Physiol 46: 929–969

Jones DP (1986) Intracellular diffusion gradients of O$_2$ and ATP. Am J Physiol 250: C663–C675

Kepler D, Decker K (1969) Studies on the mechanism of galactosamine hepatitis: accumulation of galactosamine-1-phosphate and its inhibition of UDP-glucose pyrophosphorylase. Eur J Biochem 10: 219–225

Kluytmans JH, Veenhof PR, De Zwaan A (1975) Anaerobic production of volatile fatty acids in the sea mussel *Mytilus edulis*. J Comp Physiol 104: 71–78

Kreutzer U, Siegmund B, Grieshaber MK (1985) Role of coupled substrates and alternative end products during hypoxia tolerance in marine invertebrates. Molec Physiol 8: 371–392

Mangum C, Van Winkle W (1973) Responses of aquatic invertebrates to declining oxygen conditions. Am Zool 13: 529–541

Pamatmat MM (1978) Oxygen uptake and heat production in a metabolic conformer *(Littorina irrorata)* and a metabolic regulator *(Uca pugnax)*. Mar Biol 48: 317–325

Pelster B, Bridges CR, Grieshaber MK (1988) Physiological adaptations of the intertidal rockpool teleost *Blennius pholis* to aerial exposure. Resp Physiol 71: 355–374

Plaxton WC, Storey KB (1984) Phosphorylation in vivo of red muscle pyruvate kinase from the channelled whelk, *Busycotypus canalicualtum*, in response to anoxic stress. Eur J Biochem 143: 267–272

Prosser CL (1973) Comparative animal physiology. Saunders, Philadelphia

Pörtner HO, Kreutzer U, Siegmund B, Heisler N, Grieshaber MK (1984) Metabolic adaptation of the intertidal worm *Sipunculus nudus* to functional and environmental hypoxia. Mar Biol 79: 237–247

Pörtner HO, Heisler N, Grieshaber MK (1985) Oxygen consumption and mode of energy production in the intertidal worm *Sipunculus nudus* L: definition and characterization of the critical PO$_2$ for an oxyconformer. Resp Physiol 59: 361–377

Redmond JR (1955) The respiratory function of haemocyanin in Crustacea. J Cell Comp Physiol 46: 209–247

Saz HJ (1981) Energy metabolism of parasitic helminths: adaptation to parasitism. Ann Rev Physiol 43: 323–341

Schöttler U (1977) The energy-yielding oxidation of NADH by fumarate in anaerobic mitochondria of *Tubiflex spec.* Comp Biochem Physiol 58B: 151-156

Schöttler U (1986) Weitere Untersuchungen zum anaeroben Energiestoffwechsel des Polychaeten *Arenicola marina* L. Zool Beitr NF 30: 141-152

Schöttler U, Wienhausen G, Zebe E (1983) The mode of energy production in the lugworm *Arenicola marina* at different oxygen concentrations. J Comp Physiol 149: 547-555

Schöttler U, Wienhausen G, Westermann J (1984) Anaerobic metabolism in the lugworm *Arenicola marina* L: the transition from aerobic to anaerobic metabolism. Comp Biochem Physiol 79B: 93-103

Schroff G, Schöttler U (1977) Anaerobic reduction of fumarate in the body wall musculature of *Arenicola marina*. J Comp Physiol 116: 325-336

Schultz TK, Kluytmans JH (1983) Pathway of propionate synthesis in the sea mussel *Mytilus edulis*. Comp Biochem Physiol 75B: 365-372

Teal JM, Carey FG (1967) The metabolism of marsh crabs under conditions of reduced oyxgen pressure. Physiol Zool 40: 83-90

Toulmond A, Tchernigovtzeff C (1984) Ventilation and respiratory gas exchanges of the lugworm *Arenicola marina* (L) as functions of ambient pO$_2$ (20-700 torr). Resp Physiol 57: 349-363

Ultsch GR, Ott ME, Heisler N (1980) Standard metabolic rate, critical oxygen tension, and aerobic scope for spontaneous activity of trout *(Salmo gairdneri)* and carp *(Cyprinus carpio)* in acidified water. Comp Biochem Physiol 67A: 329-335

Van Thoai N, Roche J (1964) Sur la biochimie comparée des phosphagènes et leur repartition chez les animaux. Biol Rev 39: 214-231

Walshe B (1950) The function of haemoglobin in *Chironomus plumosus* under natural conditions. J Exp Biol 27: 73-95

Wiesner RJ, Rüegg JC, Grieshaber MK (1986) The anaerobic heart: succinate formation and mechanical performance of cat papillary muscle. Exp Biol 45: 55-64

Young RE (1973) Responses to respiratory stress in relation to blood pigment affinity in *Goniopsus cruentata* (Latreille) and to a large extent in *Cardiosoma guanhumi* (Latreille). J Exp Mar Biol Ecol 11: 91-102

Metabolic and Pathological Aspects of Hypoxia in Liver Cells

H. de Groot, A. Littauer, and T. Noll

Institut für Physiologische Chemie I der Universität Düsseldorf, Moorenstraße 5, 4000 Düsseldorf, FRG

Introduction

The liver parenchymal cell, subsequently referred to as the liver cell, is among those mammalian cells which are dependent in their functions upon molecular oxygen. Hypoxia, i.e., O_2 deficiency, is present when there are deviations in the functions of the liver cell from their normal values owing to a subnormal oxygen partial pressure (PO_2). Among the oxidases and oxygenases of the liver cell, a unique role is played by cytochrome oxidase of the mitochondrial respiratory chain. The energy status and the oxidation-reduction status of the liver cell, and ultimately its viability, depend on its proper functioning. While cytochrome oxidase is characterized by an extraordinarily high affinity for O_2, other oxidases and also the oxygenases of the liver cell usually possess a significantly lower affinity for O_2. For this reason an impairment of the cytochrome oxidase activity due to O_2 deficiency, and hence cell death, only occurs under severe hypoxia. However, certain pathological cell functions may already be altered under mild hypoxia, where cytochrome oxidase activity remains unaffected. An example is the increased reductive activation of halogenated alkanes to free radicals, resulting in an increased hepatotoxicity of these compounds under hypoxia.

The present article is concerned with the mechanism of hypoxic liver cell death, as well as the role of hypoxia in the hepatotoxicity of halogenated alkanes. In connection with hypoxic cell death, the problematic nature of reoxygenation of the previously hypoxic cell will be considered. In the course of discussion of haloalkane liver injury, it will be demonstrated that despite the fact that hypoxia stimulates the deleterious effects of these compounds, the presence of small amounts of O_2 are a basic requirement for the loss of cell viability. Before treating these pathological aspects of hypoxia, the PO_2 distribution in liver, including the magnitude and course of *intercellular* and *intracellular* O_2 gradients, will be discussed, and the O_2 dependence of liver cell functions elucidated in detail.

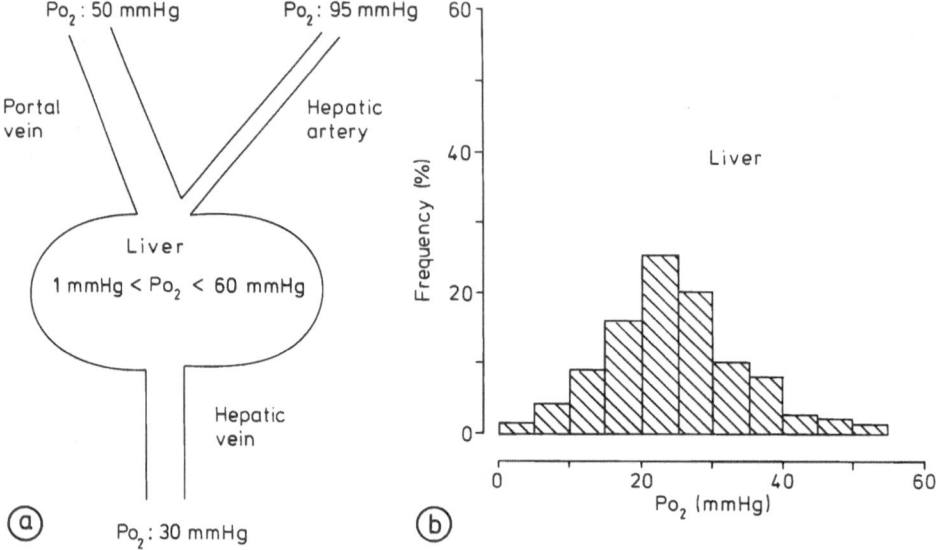

Fig. 1a, b. Supply of blood to (**a**) and PO_2 frequency distribution in the rat liver (**b**). The values of (**a**) were taken from Thurman et al. (1986a) those of (**b**) from Kessler et al. (1984)

O_2 Gradients in Liver

PO_2 Distribution in Liver

PO_2 needle electrodes and multiwire surface electrodes were used to record PO_2 frequency distribution in the livers of various experimental animals. A typical example for a PO_2 frequency distribution curve of the liver, obtained by in situ measurements, is shown in Fig. 1. The mean PO_2 is around 20 mm Hg, while the upper values reach 55–60 mm Hg and the lower values 1–5 mm Hg.

Intercellular O_2 Gradients

The liver is supplied with blood from two sources. The portal vein contributes about 75% of the total blood flow through the liver (Fig. 1). The residual 25% is derived from the hepatic artery. While the PO_2 of the arterial blood is about 95 mm Hg, the PO_2 of the portal blood ranges around 50 mm Hg (Thurman et al. 1986a). When the two blood flows intermingle in the periportal area of the liver lobule at the beginning of the liver sinusoids, the resulting PO_2 is approximately 60 mmHg. Along the liver sinusoids there are large *intercellular* O_2 gradients (Sies 1977, Ji et al. 1982). As indicated by the PO_2 frequency distribution curve, they may reach a magnitude greater than 50 mm Hg. The blood flowing through the hepatic sinusoids enters the central venule of the liver lobule and from there flows into the hepatic vein. The PO_2 of the latter blood is approximately 30 mm Hg and thus significantly higher than the mean PO_2 in liver, presumably due to inhomogeneous blood flow through the liver and inhomogeneous lengths of the liver sinusoids.

Intracellular O_2 Gradients

O_2 gradients in tissues result from the O_2 uptake of the cells. In hepatocytes, about 90% of the O_2 taken up is consumed by cytochrome oxidase of the mitochondrial respiratory chain (de Groot et al. 1985a). Due to its location within the mitochondrial compartment, O_2 gradients must exist between the extracellular space and the mitochondria. A rough approximation of the magnitude of these O_2 gradients can be derived from the O_2 dependence of the O_2 uptake of isolated hepatocytes (de Groot and Noll 1987a; see also Figs. 2, 3). The O_2 uptake rate of the isolated hepatocytes is almost independent of the PO_2 between 2-100 mmHg. At PO_2 values below 2 mmHg there is a rapid decrease in the O_2 uptake rate with decreasing PO_2, reaching a half-maximal value (P_{50} value) at a PO_2 of 0.7 mmHg. At PO_2s below the critical value of 2 mmHg the activity of mitochondrial cytochrome oxidase becomes limited by O_2 (see below). Since the cytochrome oxidase has an extraordinarily high affinity for O_2 - in mitochondria P_{50} values between 0.02 and 0.2 mmHg have been determined (Sugano et al. 1974; de Groot et al. 1985a) - it follows that at an extracellular PO_2 below 2 mmHg, the PO_2 in close vicinity to cytochrome oxidase, i.e., the intramitochondrial PO_2, must be almost zero. Thus,

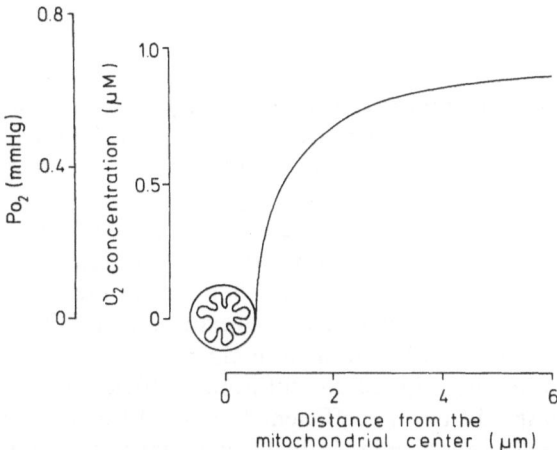

Fig. 2. O_2 gradients in the perimitochondrial zone. The PO_2, given as a function of distance r from the mitochondrial center was calculated with the equation (Boag 1969; Jones and Kennedy 1982):

$$C_r = C_\infty - \left(\frac{K\,b^2}{3\,D}\right)\frac{b}{r}$$

where C_r is the O_2 concentration in the perimitochondrial zone at a radius r from the mitochondrial center, and C_∞ the O_2 concentration of the extracellular space. Both, C_r and C_∞ should be substituted by values in μM. They can be calculated from the PO_2 by assuming 1 μM $O_2 = 0.7$ mmHg. K represents the respiratory rate of the mitochondria ($M^{-1}\,s^{-1}$), calculated from the O_2 uptake rate of the liver cells at a given PO_2 assuming that the mitochondria occupy 10% of the cell volume of 7.25 $\mu l \times 10^6$ cells^{-1} (Krebs et al. 1974; Tischler et al. 1977); b is the apparent mitochondrial radius ($= 0.6$ μm); and D is the intracellular diffusion coefficient for O_2 ($= 2.8 \times 10^{-7}$ cm^2 s^{-1}), calculated under the following assumptions: $C_r - C_\infty = 1$ μM (0.7 mmHg); $b = 0.6$ μm; $r = b$; $K = 2.3 \times 10^{-4}$ M s^{-1} as estimated for a hepatocellular O_2 uptake rate of 10 nmol $\times$ min$^{-1} \times$ 10^6 cells^{-1}

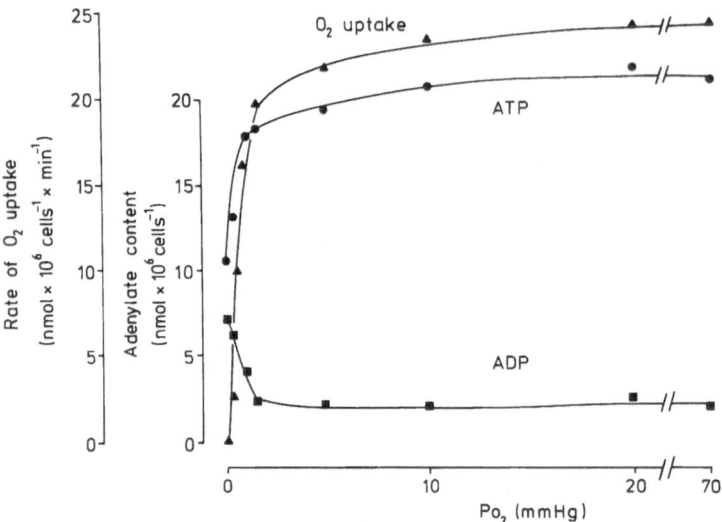

Fig. 3. O_2 dependence of the O_2 uptake rate and the ATP and ADP content of isolated hepatocytes. The incubations were performed as previously described (Noll et al. 1986). ATP and ADP were determined using enzymatic methods (Gruber et al. 1974)

at PO_2s below 2 mmHg the value of the maximal intracellular O_2 gradient between the plasma membrane and the outer mitochondrial membrane is equal to the actual value of the PO_2 by first approximation. For instance, at the PO_2 of 0.7 mmHg, where mitochondrial O_2 uptake is half-maximal, the O_2 gradient is ≤ 0.7 mmHg. From this O_2 gradient and the respective hepatocellular O_2 uptake rate an intracellular diffusion coefficient for O_2 of 2.8×10^{-7} cm^2 s^{-1} can be calculated (Fig. 2). This value is only a rough approximation of the true value. Its derivation disregards, among other things, the fact that the PO_2 at the mitochondrial surface must be above zero – it is approximately 0.1–0.2 mmHg at the extracellular PO_2 of 0.7 mmHg – and that the mitochondria may be grouped in clusters (Jones 1984). Taking both factors into consideration the value of the diffusion coefficient would increase by a factor of 5–10. With the diffusion coefficient $D = 2.8 \times 10^{-7}$ cm^2 s^{-1} and the O_2 uptake rate 10 nmol $\times$ 10^6 cells^{-1} min^{-1}, the O_2 gradient in the perimitochondrial zone can be estimated. The O_2 gradient occurs predominantly in the region immediately surrounding the mitochondria (Fig. 2), a result which is consistent with the lack of significant O_2 gradients in other parts of the cell (Jones and Mason 1978).

O_2 Dependence of Liver Cell Functions

Cytochrome Oxidase

Cytochrome oxidase is the terminal complex of the mitochondrial respiratory chain and catalyzes the reduction of O_2 to H_2O in a four-electron reaction (Nicholls 1982). The enzyme contains two molecules of haem A which occur as cytochrome a and cytochrome a_3. Cytochrome a_3 or, more precisely, its reduced form, provides the site where O_2 binds and where its reduction takes place. Unlike the remaining reactions in the mitochondrial electron-transfer chain, which are near equilibrium, the terminal reaction with O_2 is irreversible. Its rate *(v)* can be expressed by $v = k\,[a_3{}^{2+}]\,[O_2]$ (Oshino et al. 1974), where $a_3{}^{2+}$ is the reduced form of cytochrome a_3 and k is the rate constant for the reaction. According to this equation, the rate of the cytochrome a_3-O_2 reaction and, hence, the flow of electrons through the mitochondrial respiratory chain is independent of O_2 over a wide PO_2 range. Alterations in the O_2 concentration are compensated by inverse alterations in the $a_3{}^{2+}$ concentration. However, with decreasing PO_2 this compensation only works until the point is reached where all the cytochrome a_3 is present in the reduced form. From this critical PO_2 onwards, a further decrease in the PO_2 cannot be compensated and, hence, the rate of the cytochrome a_3-O_2 reaction decreases with decreasing PO_2.

About 90% of the O_2 uptake of the liver cell results from the catalytic activity of cytochrome oxidase as indicated by its inhibition by cyanide (de Groot et al. 1985 a). Thus, the decrease in the O_2 uptake rate of the isolated liver cells (Fig. 3), becoming significant at the PO_2 of 2 mmHg, apparently delineates the critical PO_2 where the cytochrome oxidase activity, and therefore also the electron transfer in the mitochondrial respiratory chain, becomes limited by O_2. In line with this contention, there is a marked fall in the ATP content of the liver cells, parallelled by an increase in their ADP content (Fig. 3). Furthermore, there is a marked increase in the NADH/NAD$^+$ concentration ratio (data not shown). Similar values for the critical PO_2 in liver have been published by other authors (Longmuir 1957; Jones and Mason 1978; Wilson et al. 1979). Besides the inhibition of cytochrome oxidase by O_2 deficiency, an increase in the glycolytic rate, accelerated by ADP, AMP, and other effectors, should also contribute to the increase in the NADH/NAD$^+$ concentration ratio. The increase in the glycolytic rate can also explain why the ATP content of the cells does not fall to zero under anaerobic conditions but remains at a constant level of $10 \text{ nmol} \times 10^6 \text{ cells}^{-1}$ (Fig. 3). A further reason for the maintenance of the ATP content at that level may be a cessation of various ATP-consuming processes (see below).

As a consequence of the limitation of cytochrome oxidase activity by O_2 there is a broad range of secondary metabolic effects. A number of processes requiring ATP come to an end. Examples include gluconeogenesis from lactate (Romero et al. 1987) and sulfate and glucuronide conjugate formation (Aw and Jones 1982). But also NADH-dependent functions of the liver cell may be impaired. For instance, NAD$^+$-dependent dehydrogenation involved in the metabolism of ethanol and other alcohols may be inhibited by the increased NADH/NAD$^+$ concentration ratio (Jones 1981).

Oxidases and Oxygenases Other than Cytochrome Oxidase

Oxidases other than cytochrome oxidase reduce O_2 in one or two electron transfer reactions to yield the superoxide anion radical (O_2^-) and hydrogen peroxide (H_2O_2), respectively. Examples include mitochondrial monoamine oxidase and peroxisomal (rat liver) urate oxidase. There are large variations in the affinity for

Table 1. P_{50} values of various oxidases and oxygenases in liver

Oxidase, oxygenase	P_{50} (mmHg)	References
Mitochondrial monoamine oxidase	100–400	Houslay and Tipton (1973) Fowler and Callingham (1978)
Peroxisomal urate oxidase	70	Sies (1977)
Microsomal amine oxidase	15–18	Poulson and Ziegler (1979)
Microsomal cytochrome P-450	3–6	Jones and Mason (1978)

P_{50}: PO_2 of half-maximal change

O_2 among the oxidases. In general their P_{50} values are at least an order of magnitude above the P_{50} value of cytochrome oxidase (Table 1). Oxygenases catalyze reactions in which one or both atoms of the O_2 molecule are inserted into the organic substrate molecule. As compared to cytochrome oxidase, relatively high P_{50} values have also been reported for this second group of enzymes requiring O_2 for their catalytic activity. As examples, P_{50} values for microsomal cytochrome P-450 and microsomal amine oxidase are included in Table 1. Cytochrome P-450 stands for a group of isoenzymes catalyzing the monooxygenation of a variety of lipophilic drugs. Microsomal amine oxidase is an enzyme containing FAD that catalyzes the oxygenation of numerous amines and hydrazines.

Comparison of the P_{50} values given in Table 1 with the PO_2 distribution in liver (Fig. 1) reveals that the catalytic activities of almost all oxygenases and oxidases, except for cytochrome oxidase, may become limited by O_2 even at physiological PO_2 levels. Thus, while severe hypoxia is necessary for the restriction of cytochrome oxidase activity by O_2, mild hypoxia may already lead to substantial decreases in the activity of the other oxidases and oxygenases of the liver.

Hypoxic Liver Cell Death and Reoxygenation Injury

Hypoxic Cell Death

Liver cell necrosis due to O_2 deficiency occurs following circulatory disorders caused by severe heart insufficiency or shock. It may also play a role in ethanol liver injury (Israel et al. 1975). Due to intercellular O_2 gradients (see above) centrilobularily located cells are most vulnerable to hypoxia. Thus, it is not surprising that centrilobular necrosis is the typical pathological finding in the liver injuries mentioned above.

It is generally accepted that the depression of the ATP synthesis rate due to O_2 limitation of cytochrome oxidase activity is the decisive functional lesion responsible for hypoxic cell death (Farber et al. 1981; Hochachka 1986). ATP depletion is among the first alterations detectable in the hypoxic liver cell (see previous section). In addition, in the isolated perfused liver it has been demonstrated that fructose has a protective effect against hypoxic cell death, presumably mediated by an increased glycolytic ATP production (Anundi et al. 1987).

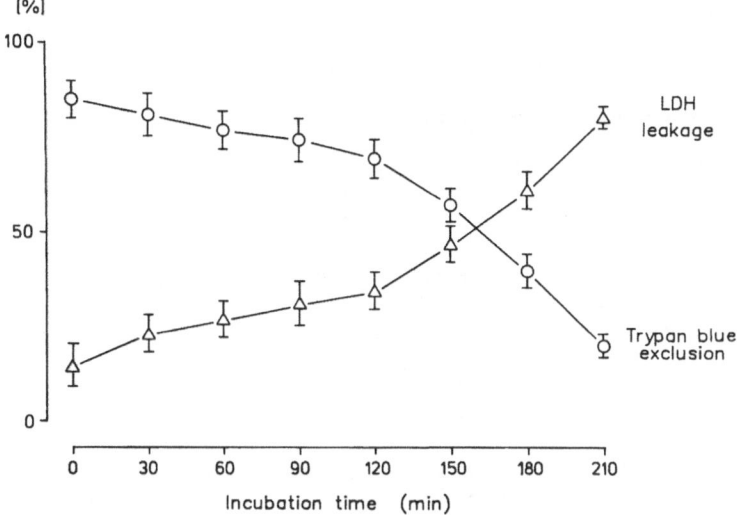

Fig. 4. Trypan blue exclusion and lactate dehydrogenase (LDH) leakage during incubation of isolated hepatocytes from fed rats under anaerobic conditions. The incubations were performed as previously described (Littauer et al. 1988)

In Fig. 4 an anaerobic incubation of isolated hepatocytes is shown. During the first 2 h there was only a slight decrease in trypan blue exclusion and a small increase in lactate dehydrogenase leakage. After that, however, marked changes in both indicators of severe cell injury occurred, indicating an almost complete loss of viable cells after 3.5 h. Thus, a distinct period of time elapses between the early impairment of the energy status developing within minutes and the loss of cell viability becoming visible after about 2 h. As yet, it is uncertain which processes within the cell connect the two events.

A disturbance of the ion homeostasis resulting from the impaired energy status of the cell is presumed to be the critical further step towards cell death following ATP depletion. Although an increase in intracellular sodium ions and a decrease in intracellular potassium ions, possibly due to the high K_m for ATP of the plasmalemmal Na^+-K^+-ATPase, is widely recognized, the increase of cytosolic free calcium ions is regarded as especially critical (Jennings et al. 1975; Farber et al. 1981; Nayler 1983; Snowdowne et al. 1985). A rise in the cytosolic Ca^{2+} level may be evoked by several mechanisms. It may be mediated by an increased influx from the extracellular space, due either to depressed activity of ATP-dependent transporters or to reversal of the Na^+-Ca^{2+}-antiporter as a consequence of the elevated Na^+ levels in the cell. It may also result from a disturbed sequestration of Ca^{2+} by the endoplasmic reticulum and the mitochondria. With regard to cell death, one of the decisive subsequent steps is hypothesized to be represented by the intracellular Ca^{2+}-dependent activation of phospholipase A (Fig. 5). In line with this assumption there is a decrease in cellular phospholipids during hypoxia (Farber and Young 1981), resulting in an increase in the cholesterol-to-phospholipid ratio (Petrovich et al. 1984). Concomitantly, an increase in decay products of phospholip-

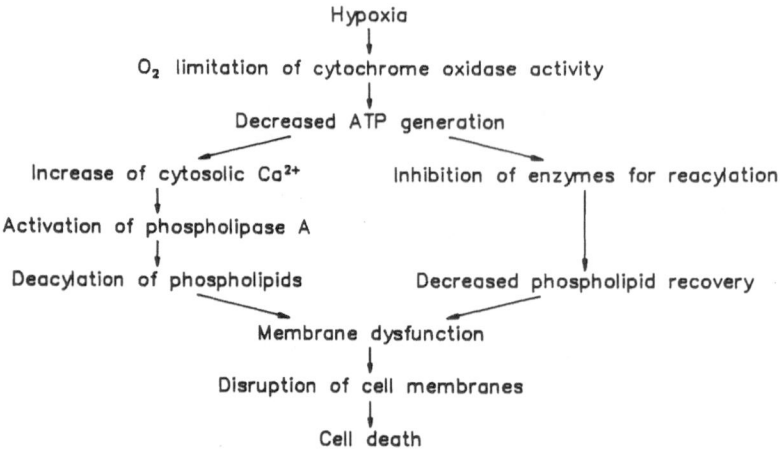

Fig. 5. Proposed mechanism of hypoxic cell death involving the degradation of membrane phospholipids (Farber and Young 1981; Das et al. 1986)

ids occurs (Farber and Young 1981). Furthermore, a decreased reacylation of the decay products of phospholipids may play a role, as has been observed in heart muscle cells (Das et al. 1986). These processes are assumed to be accompanied by alterations in the physical properties of the membranes and, thus, in changes in the lipid environment of membrane-bound enzymes. The resulting decreased enzymatic activities may evoke a further disturbance of the ion homeostasis. Ultimately, cell death may occur by disruption of cellular membranes.

Although the pathway of hypoxic liver cell necrosis as outlined above is supported by several lines of evidence, there are serious arguments against it. For instance, the assumption that an increase in the cytosolic Ca^{2+} is the crucial event for the following processes, such as phospholipid degradation, may be invalid. Lemasters et al. (1987) were unable to detect such an increase in hepatocytes, and the results of experiments in tissues other than liver are still controversial (Allen and Orchard 1983; Cheung et al. 1986; Cobbold et al. 1985; Snowdowne et al. 1985). Likewise, although extracellular Ca^{2+} is supposed to influence intracellular Ca^{2+} levels, the effect of removal of extracellular Ca^{2+} on hypoxic cell death is the subject of controversial discussion (Cheung et al. 1982; Okuno et al. 1983). Hence, alternative pathways to irreversible cell injury during hypoxia must be considered as well. For example, it has been suggested that ATP depletion results in a failure of the cytoskeleton to maintain cell shape and volume (Lemasters et al. 1982).

Reoxygenation Injury

Severe hypoxia will ultimately produce liver cell death if it lasts long enough. Thus, the flow of blood and, with it, the supply of O_2 has to be restored before extended cell death occurs. This problem area, for example, is of great importance for liver transplantation.

One would expect the restoration of normal PO_2 to stop damage occurring during the hypoxic period and to allow recovery of reversibly injured cells. Indeed, O_2 replacement after an anoxic period of up to 2 h showed a salutary effect in isolated hepatocytes. However, when reoxygenation was established after an anaerobic period of 2 h or longer, O_2 did not cause the expected improving effect. On the contrary, cell death accelerated as compared to the anaerobic control incubation (preliminary results). At present, the reason for this oxygen paradox is still unknown.

It was suggested that O_2-derived free radicals play the decisive role in reoxygenation injury (McCord 1985). According to this hypothesis, during the ischemic period ATP is degraded into hypoxanthine and in parallel xanthine dehydrogenase is converted to xanthine oxidase due to ATP depletion followed by Ca^{2+} influx (Roy and McCord 1983). With the reavailability of O_2, xanthine oxidase catalyzes the oxidation of hypoxanthine to yield uric acid and the reduced O_2 species O_2^- and H_2O_2. Both may react in a transition metal-catalyzed, Haber-Weiss reaction to yield the highly reactive hydroxyl radical ($OH\cdot$, Parks and Granger 1983). Attack by the reactive O_2-derived free radicals on essential cellular compounds is considered to be responsible for cell death. However, there are no data available with regard to the involvement of free radicals in reoxygenation injury in liver. In recent reoxygenation studies in isolated hepatocytes, we found no evidence for an attack by free radicals on cellular membranes. Furthermore, in tissues other than liver, the effect of free radical scavenging substances, such as superoxide dismutase, dimethylsulfoxide and catalase, and the role of allopurinol, a specific inhibitor of xanthine oxidase, are being controversially discussed (Shlafer et al. 1982; Stewart et al. 1982; Parks et al. 1983; Parks and Granger 1983; Moorhouse et al. 1987; Kehrer et al. 1987). Moreover, the assumption that xanthine dehydrogenase converts to xanthine oxidase may not apply (Kehrer et al. 1987). Thus, the involvement of free radicals in reoxygenation injury remains doubtful.

Another hypothesis is based on the assumption that a disruption of the cytoskeleton occurs by reoxygenation (Lemasters et al. 1983). After O_2 has been introduced to the previously hypoxic liver, the authors observed a shrinkage of hepatocytes and a release of blebs. In this case, the elevation of the cellular ATP level during reoxygenation may lead to a disruption of the cytoskeleton in a similar manner as described for the contracture in myocard cells (Kloner et al. 1974).

An experimental approach to elucidate the mechanism of the oxygen paradox may possibly be provided by the finding that the aggravation effect of O_2 sets in just when severe damage of the liver cells occurs under anaerobic conditions (see above). This implies that the key to the understanding of the oxygen paradox may lie in those alterations already occurring during the hypoxic period.

Hypoxia and Haloalkane Liver Injury

Studies of the mechanisms of liver injury have led to an accumulation of evidence that hypoxia can play a crucial role during the perpetuation of those pathological alterations caused by chemical compounds foreign to the mammalian organism. Examples include ethanol (for review, see Thurman et al. 1986b), which obviously still remains one of the most important hepatotoxins in the public health, and halogenated hydrocarbons (for review, see Brattin et al. 1985; Cheeseman et al. 1985; de Groot and Noll 1986), which represent a widespread class of synthetic chemicals employed in industry, households, agriculture, and medicine as solvents (carbon tetrachloride), pesticides (DDT, 1,1,1-trichloro-2,2-bis (p-chlorophenyl) ethane); lindane, (γ-hexachlorocyclohexane), and inhalation anesthetics (halothane, 2-bromo-2-chloro-1,1,1-trifluoroethane). Therefore, there are many different circumstances under which exposure may occur, including the inhalation of fumes during industrial accidents, inadvertent ingestion of solvents, and inhalation of anesthetics during surgery.

Liver injury developing upon exposure to haloalkanes is characterized by steatosis and/or necrosis of those liver cells localized in the centrilobular region of the liver lobule. However, individual haloalkanes can considerably differ in the degree and time course of manifestation of the pathological stages (cf. Zimmerman 1978). Studies mainly performed with the model hepatotoxin carbon tetrachloride (CCl_4) have confirmed that the initial steps leading to liver cell injuries require a biotransformation of the foreign compound to a reactive metabolite catalyzed by the microsomal cytochrome P-450 enzyme system (Recknagel and Ghoshal 1966; Slater 1966; Reynolds and Moslen 1980; de Groot and Noll 1986). Those isoenzymes of cytochrome P-450 induced by pretreatment of experimental animals with phenobarbital are especially effective in this respect (Noguchi et al. 1982). The biotransformation represents an NADPH-dependent reductive dehalogenation of CCl_4 leading to the correspondent carbon-centered trichloromethane radical ($CCl_3 \cdot$). This step occurs via an one-electron reduction of the haloalkane and requires an interaction with the haem-iron of cytochrome P-450, a location where O_2 is normally activated during the monooxygenase cycle. The reduction equivalents are supplied via the cytochrome P-450 reductase. Several lines of evidence point to the formation of an intermediate haloalkane-haem-iron complex during the metabolic activation of CCl_4 and other homologous compounds (Ullrich 1979, Ahr et al. 1980; Mansuy and Battioni 1985). Due to the competition between O_2 and the haloalkanes for the electrons and the common site of activation, generation of haloalkane radicals increases with decreasing PO_2, and is thus maximal under anaerobic conditions.

Once a reactive $CCl_3 \cdot$ is generated, it can attack certain attractive sites of cellular macromolecules. As a consequence, $CCl_3 \cdot$ radicals bind to microsomal proteins and lipids spatially confined near the location of their formation (Reynolds and Yee 1967; Uehleke et al. 1973; Noguchi et al. 1982). In the presence of O_2, the interaction of CCl_4-derived radicals ($CCl_3 \cdot$, $CCl_3O_2 \cdot$, and $CCl_3O \cdot$) with the unsaturated fatty acids of membranal phospholipids can initiate a self-propagating lipid peroxyl free-radical-carried chain reaction which ultimately leads to the disruption of cellular membranes (for a recent review, see de Groot and Noll 1987 b).

As shown by Reynolds and Yee (1967), Uehleke et al (1973), and Wolf et al. (1980), reductive activation of CCl_4 accompanied by covalent binding of CCl_4 metabolites to the microsomal protein and lipid fraction is drastically enhanced under anaerobic conditions. These observations, together with the in vivo results that hypoxia increases haloalkane-induced liver cell injury, as indicated by an enlargement of the areas of centrilobular cell necrosis (Strubelt and Breining 1980; Shen et al. 1982), have led to the assumption that covalent binding of CCl_4 metabolites to cellular macromolecules is a causative of the haloalkane-mediated liver cell injury. As shown in microsomes, however, except in the case of the cytochrome P-450 enzyme function (de Groot and Haas 1980), covalent binding of haloalkane metabolites to microsomal proteins and lipids did not cause impairments of enzyme or membrane functions per se (de Groot et al. 1985b, 1986). These results have been verified by incubations of CCl_4-supplemented isolated hepatocytes under anoxia (Littauer et al. 1988). Although the generation of $CCl_3 \cdot$ radicals should be maximal under these conditions, no significant effect of CCl_4 on cell viability could be detected (Fig. 7).

In contrast, incubations of isolated microsomes and hepatocytes under defined steady-state O_2 conditions by use of an especially designed oxystat system (Noll et al. 1986, 1987) revealed that the CCl_4-induced lipid peroxidation is significantly enhanced when the steady-state PO_2 level is shifted from values around 70 mmHg to values around 7 mmHg (Noll and de Groot 1984; Littauer et al. 1988; Noll et al. 1987, Fig. 6) and is accompanied by drastic alterations of microsomal enzyme activities, like UDP glucuronyl transferase, nucleoside diphosphatase, and glucose-6-phosphatase, as well as loss of the barrier function of the microsomal membrane (de Groot et al. 1985b, 1986; de Groot and Noll 1986). Evaluation of CCl_4-mediated liver cell injury by trypan blue uptake and lactate dehydrogenase leakage revealed that the presence of O_2 is a prerequisite for the initiation of the hepatocellular injury caused by this compound, and that liver cell injury was drastically increased at those PO_2 levels which were optimal for CCl_4-induced lipid peroxidation (Fig. 7; for more details, see Littauer et al. 1987).

The unusual O_2 dependence of CCl_4-induced lipid peroxidation can be explained if it is taken into account that, on the one hand, the generation of the

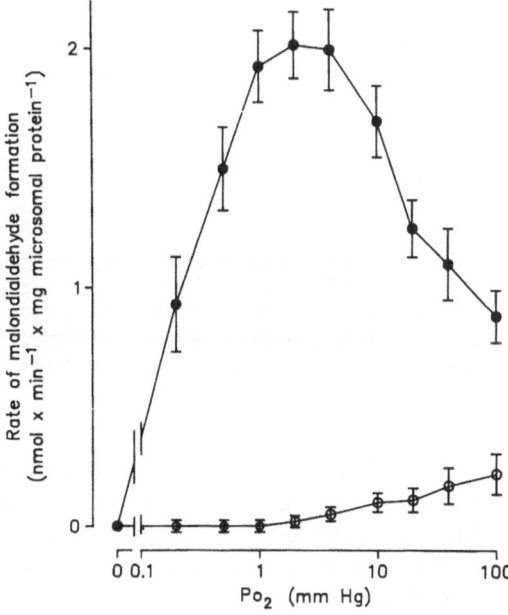

Fig. 6. O_2 dependence of CCl_4-induced lipid peroxidation as indicated by malondialdehyde formation in rat liver microsomes. For experimental details, see Noll and de Groot (1984)

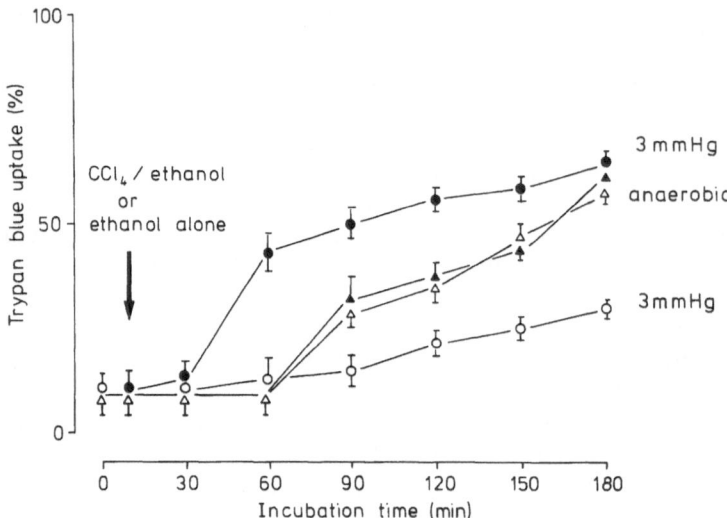

Fig. 7. O_2 dependence of the CCl_4-induced loss of cell viability in rat hepatocytes. ● CCl_4/ethanol, ○ ethanol alone. For experimental details, see Littauer et al. (1988)

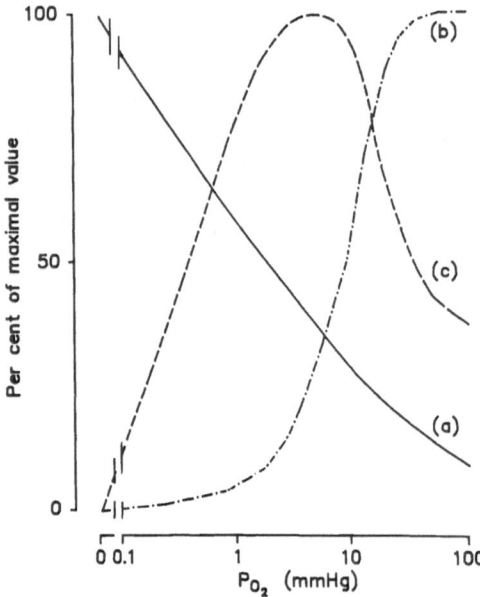

Fig. 8. O_2 dependence of the generation of haloalkane free radicals *(a)*, peroxidative breakdown of unsaturated fatty acids of membranal phospholipids *(b)*, and haloalkane-induced peroxidative membrane damage *(c)*

$CCl_3 \cdot$ radicals increases with decreasing PO_2 while, on the other hand, O_2 is a key reactant of the lipid peroxyl free-radical-carried reactions of the membranal lipids (see de Groot and Noll 1987b; Fig. 8). Since both processes are coupled in series, it becomes obvious that due to the opposite roles of O_2 in each step, CCl_4-induced lipid peroxidation occurs at PO_2 levels which are low enough to permit the reductive activation to the $CCl_3 \cdot$ radicals and high enough to insure the propagation of lipid peroxidation.

These findings demonstrate that lipid peroxidation rather than covalent binding of reactive CCl_4 metabolites to cellular macromolecules plays a crucial role in the CCl_4-mediated liver cell damage. As shown for other halomethanes like $CBrCl_3$, $CHBr_3$ (de Groot and Noll 1985b, 1986), and haloethanes like the common inhalation anesthetic halothane ($CF_3CHClBr$; de Groot and Noll 1983, 1984, 1985a, the initiation of lipid peroxidation, occurring preferentially or exclusively at low PO_2 levels, accompanied by losses of membrane-linked enzyme functions, appears to be a common feature of haloalkane-mediated pathological processes within the hepatocyte.

Since lipid peroxidation induced by haloalkanes is drastically enhanced at PO_2 levels between 1 and 20 mmHg, as compared to PO_2 levels around 70 mmHg, and since lipid peroxidation coincides with the losses of cellular membrane and enzyme functions as well as with the loss of cell viability, it appears to be obvious that the spatial distribution of O_2 within the liver lobule plays a decisive role for the spatial distribution of the haloalkane-induced cell necrosis. Since hypoxia, as caused by a temporal malperfusion of the liver, leads to an enlargement of those areas with cells exposed to low PO_2 levels, it concomitantly increases the vulnerability of the liver upon exposure to haloalkanes.

Acknowledgements. We thank Professor H. Sies for his helpful comments and suggestions during the course of these studies. Work at the authors' laboratory was generously supported by Ministerium für Wissenschaft und Forschung, Nordrhein-Westfalen and the Deutsche Forschungsgemeinschaft, Schwerpunktprogramm "Mechanismen toxischer Wirkungen von Fremdstoffen"

References

Ahr HJ, King LJ, Nastainczyk W, Ullrich V (1980) The mechanism of chloroform and carbon monoxide formation from carbon tetrachloride by microsomal cytochrome P-450. Biochem Pharmacol 29: 2855–2861

Allen DG, Orchard DG (1983) Intracellular calcium concentration during hypoxia and metabolic inhibition in mammalian ventricular muscle. J Physiol (Lond) 339: 107–122

Anundi I, King J, Owen DA, Schneider H, Lemasters JJ, Thurman RG (1987) Fructose prevents hypoxic cell death in liver. Am J Physiol 253: G390–G369

Aw TY, Jones DP (1982) Secondary bioenergetic hypoxia. Inhibition of sulfation and glucuronidation reactions in isolated hepatocytes at low O_2 concentration. J Biol Chem 257: 8997–9004

Boag JW (1969) Oxygen diffusion and oxygen depletion problems in radiobiology. In: Ebert M, Howard A (eds) Current topics in radiation research. North-Holland, Amsterdam, pp 141–193

Brattin WJ, Glende EA, Recknagel RO (1985) Pathological mechanisms in carbon tetrachloride hepatotoxicity. J Free Radic Biol Med 1: 27–38

Cheeseman KH, Albano EF, Tomasi A, Slater TF (1985) Biochemical studies of the metabolic activation of halogenated alkanes. Environ Health Perspect 64: 85–101

Cheung JY, Thompson IG, Bonventre JV (1982) Effects of extracellular calcium removal and anoxia on isolated rat myocytes. Am J Physiol 243: C184–C190

Cheung JY, Leaf A, Bonventre JV (1986) Mitochondrial function and intracellular calcium in anoxic cardiac myocytes. Am J Physiol 250: C18–C25

Cobbold PH, Bourne PK, Cuthbertson KSR (1985) Evidence from aequorin for injury of metabolically inhibited myocytes independently of free Ca^{2+}. In: Spieckermann PG, Piper HM (eds) Isolated adult cardiac myocytes. Steinkopff, Darmstadt, pp 155–158

Das DK, Engelman RM, Rousou JA, Breyer RH, Otani H, Lemeshow S (1986) Role of membrane phospholipids in myocardial injury induced by ischemia and reperfusion. Am J Physiol 251: H71–H79

de Groot H, Haas W (1980) Oxygen-independent damage of cytochrome P-450 by CCl_4-metabolites in hepatic microsomes. FEBS Lett 115: 153–256

de Groot H, Noll T (1983) Halothane hepatotoxicity: relation between metabolic activation, hypoxia, covalent binding, lipid peroxidation and liver cell damage. Hepatology 3: 601–606

de Groot H, Noll T (1984) The crucial role of hypoxia in halothane-induced lipid peroxidation. Biochem Biophys Res Commun 119: 139–143

de Groot H, Noll T (1985a) Halothane-induced lipid peroxidation and glucose-6-phosphatase inactivation in microsomes under hypoxic conditions. Anesthesiology 62: 44–48

de Groot H, Noll T (1985b) Haloalkane free radicals and lipid peroxidation under low steady-state oxygen partial pressures. In: Poli G, Cheeseman KH, Dianzani MU, Slater TF (eds) Free radicals in liver injury. IRL, Oxford, pp 185–189

de Groot H, Noll T (1986) The crucial role of low steady-state oxygen partial pressures in haloalkane free-radical-mediated lipid peroxidation. Possible implications in haloalkane liver injury. Biochem Pharmacol 35: 15–19

de Groot H, Noll T (1987a) Oxygen gradients: the problem of hypoxia. Biochem Soc Trans 15: 363–365

de Groot H, Noll T (1987b) The role of physiological oxygen partial pressures in lipid peroxidation. Theoretical considerations and experimental evidence. Chem Phys Lipids 44: 209–226

de Groot H, Noll T, Sies H (1985a) Oxygen dependence and subcellular partitioning of hepatic menadione-mediated oxygen uptake. Studies with isolated hepatocytes, mitochondria, and microsomes in an oxystat system. Arch Biochem Biophys 243: 556–562

de Groot H, Noll T, Tölle T (1985b) Loss of latent activity of liver microsomal membrane enzymes evoked by lipid peroxidation. Studies of nucleoside diphosphatase, glucose-6-phosphatase, and UDP glucuronyl transferase. Biochim Biophys Acta 815: 91–96

de Groot H, Noll T, Rymsa B (1986) Alterations of the microsomal glucose-6-phosphatase system evoked by ferrous iron- and haloalkane free-radical-mediated lipid peroxidation. Biochim Biophys Acta 881: 350–355

Farber JL, Young EE (1981) Accelerated phospholipid degradation in anoxic rat hepatocytes. Arch Biochem Biophys 211: 312–320

Farber JL, Chien KR, Mittnacht S (1981) The pathogenesis of irreversible cell injury in ischemia. Am J Pathol 102: 271–281

Fowler CJ, Callingham BA (1978) Substrate-selective activation of rat liver mitochondrial monoamine oxidase by oxygen. Biochem Pharmacol 27: 1995–2000

Hochachka PW (1986) Defence strategies against hypoxia and hypothermia. Science 231: 234–241

Houslay MD, Tipton KF (1973) The reaction pathway of membrane-bound rat liver mitochondrial monoamine oxidase. Biochem J 135: 735–750

Israel Y, Kalant H, Orrego H, Khauna JM, Videla L, Phillips JM (1975) Experimental alcohol-induced hepatic necrosis: suppression by propylthiouracil. Proc Natl Acad Sci USA 72: 1137–1141

Jennings RB, Ganote CE, Reimer K (1975) Ischemic tissue injury. Am J Pathol 81: 179–198

Ji S, Lemasters JJ, Christenson V, Thurman RG (1982) Periportal and pericentral pyridine nucleotide fluorescence from the surface of the perfused liver: evaluation of the hypothesis that chronic treatment with ethanol produces pericentral hypoxia. Proc Natl Acad Sci USA 79: 5415–5419

Jones DP (1981) Hypoxia and drug metabolism. Biochem Pharmacol 30: 1019–1023

Jones DP (1984) Effect of mitochondrial clustering on O_2 supply in hepatocytes. Am J Physiol 247: C83–C89

Jones DP, Kennedy FG (1982) Intracellular oxygen supply during hypoxia. Am J Physiol 243: C247–C253

Jones DP, Mason HS (1978) Gradients of oxygen concentration in hepatocytes. J Biol Chem 253: 4874–4880

Kehrer JP, Piper HM, Sies H (1987) Xanthine oxidase is not responsible for reoxygenation injury in isolated-perfused rat heart. Free Radic Res Comm 3: 69–78

Kessler M, Höper J, Harrison DK, Skolasinska K, Klövekorn WP, Sebening F, Volkholz HJ, Beier I, Kernbach C, Rettig V, Richter H (1984) Tissue oxygen supply under normal and pathological conditions. In: Lübbers DW, Acker H, Leniger-Follert E, Goldstick TK (eds) Oxygen transport to tissue-V. Plenum, New York, pp 69–80

Kloner RA, Ganote CE, Whalen DA Jr, Jennings RB (1974) Effect of a transient period of ischemia on myocardial cells. II. Fine structure during the first few minutes of reflow. Am J Pathol 74: 399–422

Krebs HA, Cornell NW, Lund P, Hems R (1974) Isolated liver cells as experimental material. In: Lundquist R, Tygstrup N (eds) Regulation of hepatic metabolism. Academic, New York, pp 726–750

Lemasters JJ, Stemkowski CJ, Ji S, Thurman RG (1982) Liver structure and function in hypoxia. In: Wauquier A et al (eds) Protection of tissues against hypoxia. Elsevier, Amsterdam, pp 15–30

Lemasters JJ, Stemkowski CJ, Ji S, Thurman RG (1983) Cell surface changes and enzyme release during hypoxia and reoxygenation in the isolated, perfused rat liver. J Cell Biol 97: 778–786

Lemasters JJ, DiGuiseppi J, Nieminen A-L, Herman B (1987) Blebbing, free Ca^{2+} and mitochondrial membrane potential preceding cell death in hepatocytes. Nature 325: 78–81

Littauer A, Hugo-Wissemann D, Noll T, de Groot H (1988) Molecular oxygen is essential for carbon tetrachloride-mediated loss of cell viability in isolated hepatocytes. Life Sci Adv (in press)

Longmuir IS (1957) Respiration rate of rat liver cells at low oxygen concentrations. Biochem J 65: 378–382

Mansuy D, Battioni P (1985) Particular ability of cytochrome P-450 to form reactive intermediates and metabolites. In: Siest G (ed) Drug metabolism, molecular approaches and pharmalogical implications. Pergamon, Oxford, pp 195–203

McCord JM (1985) Oxygen-derived free radicals in postischemic tissue injury. N Engl J Med 312: 159–163

Moorhouse PC, Grootveld M, Halliwell B, Quinlan JG, Gutteridge JMC (1987) Allopurinol and oxypurinol are hydroxyl radical scavengers. FEBS Lett 213: 23–28

Nayler WG (1983) Calcium and cell death. Eur Heart J. 4: 33–41

Nicholls DG (1982) Bioenergetics. Academic, London

Noguchi T, Fong K-L, Lai EK, Alexander SS, King MM, Olson L, Poyer JL, McCay PB (1982) Specificity of a phenobarbital-induced cytochrome P-450 for metabolism of carbon tetrachloride to the trichloromethyl radical. Biochem Pharmacol 31: 615–624

Noll T, de Groot H (1984) The critical steady-state hypoxic conditions in carbon tetrachloride-induced lipid peroxidation in rat liver microsomes. Biochim Biophys Acta 795: 356–362

Noll T, de Groot H, Wissemann P (1986) A computer-supported oxystat system maintaining steady-state oxygen partial pressures and simultaneously monitoring oxygen uptake in biological systems. Biochem J 236: 765–769

Noll T, Hugo-Wissemann D, Littauer A, de Sagara RM, de Groot H (1987) The decisive oxygen partial pressure-levels in haloalkane-mediated liver cell injury. Free Radic Res Comm 3: 293–298

Okuno F, Orrego H, Israel Y (1983) Calcium requirement for anoxic liver cell injury. Res Commun Chem Pathol Pharmacol 39: 437–444

Oshino N, Sugano T, Oshino R, Chance B (1974) Mitochondrial function under hypoxic conditions: the steady states of cytochrome $a + a_3$ and their relation to mitochondrial energy states. Biochim Biophys Acta 368: 298–310

Parks DA, Granger DN (1983) Ischemia-induced vascular changes: role of xanthine oxidase and hydroxyl radicals. Am J Physiol 245: G285–G289

Parks DA, Bulkley GB, Granger DN (1983) Role of oxygen-derived free radicals in digestive tract diseases. Surgery 94: 415–422

Petrovich DR, Finkelstein S, Waring AJ, Farber JL (1984) Liver ischemia increases the molecular order of microsomal membranes by increasing the cholesterol-to-phospholipid ratio. J Biol Chem 259: 13217–13223

Poulson LL, Ziegler DM (1979) The liver microsomal FAD-containing monooxygenase. Spectral characterization and kinetic studies. J Biol Chem 254: 6449–6455

Recknagel RO, Ghoshal AK (1966) Lipoperoxidation as a vector in carbon tetrachloride hepatotoxicity. Lab Invest 15: 132–146

Reynolds ES, Yee AG (1967) Liver parenchymal cell injury. V. Relationships between patterns of chloromethane-C^{14} incorporation into constituents of liver in vivo and cellular injury. Lab Invest 16: 591–603

Reynolds ES, Moslen MT (1980) Free-radical damage in liver. In: Pryor WA (ed) Free radicals in biology, vol IV. Academic, New York, pp 49–94

Romero FJ, Pallardo FV, Bolinches R, Saez GT, Noll T, de Groot H (1987) Dependence of hepatic gluconeogenesis on oxygen partial pressure. Inhibitory effects of halothane. J Appl Physiol 63: 1776–1780

Roy RS, McCord JM (1983) Superoxide and ischemia: conversion of xanthine dehydrogenase to xanthine oxidase. In: Greenwald RA, Cohen G (eds) Oxy radicals and their scavenger systems. Elsevier, Amsterdam, pp 145–153

Shen ES, Garry VF, Anders MW (1982) Effect of hypoxia on carbon tetrachloride hepatotoxicity. Biochem Pharmacol 31: 3787–3793

Shlafer M, Kane PF, Wiggins VY, Kirsh MM (1982) Possible role for cytotoxic oxygen metabolites in the pathogenesis of cardiac ischemic injury. Circulation 66: I85–I92

Sies H (1977) Oxygen gradients during hypoxic steady states in liver. Urate oxidase and cytochrome oxidase as intracellular O_2 indicators. Biol Chem Hoppe Seyler 358: 1021–1032

Slater TF (1966) Necrogenic action of carbon tetrachloride in the rat: a speculative mechanism based on activation. Nature 209: 36–40

Snowdowne KW, Freudenrich CC, Borle AB (1985) The effects of anoxia on cytosolic free calcium, calcium fluxes, and cellular ATP levels in cultured kidney cells. J Biol Chem 260: 11619–11626

Stewart JR, Blackwell WH, Crute SL, Loughlin V, Hess ML, Greenfield LJ (1982) Prevention of myocardial ischemia/reperfusion injury with oxygen free-radical scavengers. Surg Forum 33: 317–320

Strubelt O, Breining H (1980) Influence of hypoxia on the hepatotoxic effects of carbon tetrachloride, paracetamol, allyl alcohol, bromobenzene and thioacetamide. Toxicol Lett 6: 109–113

Sugano T, Oshino N, Chance B (1974) Mitochondrial functions under hypoxic conditions. The steady states of cytochrome c reduction and of energy metabolism. Biochim Biophys Acta 146: 340–358

Thurman RG, Ji J, Lemasters JJ (1986a) Lobular oxygen gradients: possible role in alcohol-induced hepatotoxicity. In: Thurman RG, Kauffman FC, Jungermann K (eds) Regulation of hepatic metabolism. Intra- and intercellular compartmentation. Plenum, New York, pp 293–320

Thurman RG, Kauffman FC, Baron J (1986b) Biotransformation and zonal toxicity. In: Thurman RG, Kauffman FC, Jungermann K (eds) Regulation of hepatic metabolism. Intra- and intercellular compartmentation. Plenum, New York, pp 321–382

Tischler ME, Hecht P, Williamson JR (1977) Determination of mitochondrial/cytosolic metabolite gradients in isolated rat liver cells by cell disruption. Arch Biochem Biophys 181: 278–292

Uehleke H, Hellmer KH, Tabarelli S (1973) Binding of ^{14}C-carbon tetrachloride to microsomal protein in vitro and formation of $CHCl_3$ by reduced liver microsomes. Xenobiotica 3: 1–11

Ullrich V (1979) Cytochrome P-450 and biological hydroxylation reactions. Top Curr Chem 83: 67–104

Wilson DF, Owen CS, Erecinska M (1979) Quantitative dependence of mitochondrial oxidative phosphorylation on oxygen concentration: a mathematical model. Arch Biochem Biophys 195: 494–504

Wolf CR, Harrelson WG, Nastainczyk WM, Philpot RM, Kalyanaraman R, Mason RP (1980) Metabolism of carbon tetrachloride in hepatic microsomes and reconstituted monooxygenase systems and its relationship to lipid peroxidation. Mol Pharmacol 18: 553–558

Zimmerman HJ (1978) Hepatotoxicity. The adverse effects of drugs and other chemicals on the liver. Appleton-Century-Crofts, New York

Possible Mechanisms of O_2 Sensing in Different Cell Types

H. Acker

Max-Planck-Institut für Systemphysiologie, Rheinlanddamm 201, 4600 Dortmund 1, FRG

The human organism needs a continuous supply of oxygen to maintain its specific cell functions under different working conditions, for example, a suitable heart rate, a well-modulated excretory capacity of the kidneys, or an appropriate activity of the central nervous system. Oxygen is a vital substrate for the human body, for no metabolic pathway can generate this substance for the organism. It seems to be, therefore, reasonable to assume that the different organs have developed mechanisms to guarantee a constant oxygen supply. That means, besides its role as electron receptor in the respiratory chain, oxygen is involved as a signal in different physiological reactions. This process is defined as oxygen sensing and describes why cells are able to respond to PO_2 changes by altering the corresponding metabolic and membrane properties in order to regulate cell-specific activities and to maintain a regular function. To elucidate the mechanisms of the O_2 sensing process this article first describes this phenomenon in different biological systems to continue with a description of its basic properties. The article will be summarized by a model, which may help to stimulate further discussion about O_2 sensing.

Phenomenon

The following biological systems have been used to demonstrate the expression of the oxygen sensing process:

1. It is well documented in the literature, that organs like brain or heart can increase their local perfusion under hypoxic conditions (for review see [2]). This is also very well demonstrated in the carotid body with a PO_2-sensitive local flow regulation [1]. It seems that the capillary network in the carotid body is capable of detecting PO_2 changes, since, as shown in Fig. 1, in hyperoxia a predominance of vessels with small diameters could be observed, whereas under low oxygen pressure the vessels were dilated. The type of vessels, which have been evaluated, comprises small arterioles as well as capillaries [33]. It is reasonable, therefore, to suggest that endothelial cells in the carotid body are PO_2 sensitive and contribute to local flow changes under hypoxic sensations.

Percentage vascular volume/total volume

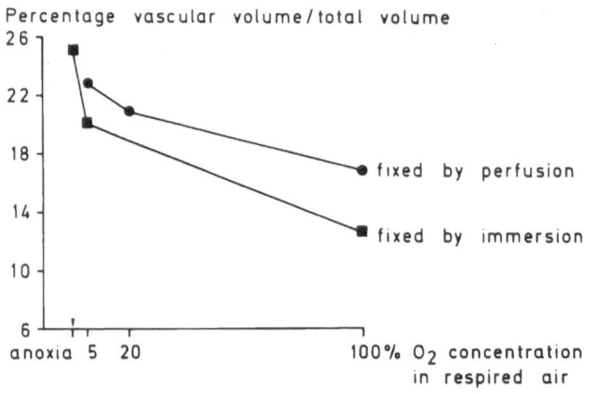

Fig. 1. Dependence of the vascular volume in the cat carotid body on the oxygen concentration in the respired air of the animal. The carotid body was fixed either by immersion or perfusion with Bouin solution. It is to be seen that the portion of the vascular volume on the total volume of the organ is increasing with lowering the O_2 concentration meaning a vasodilatation (Seidl et al. [33])

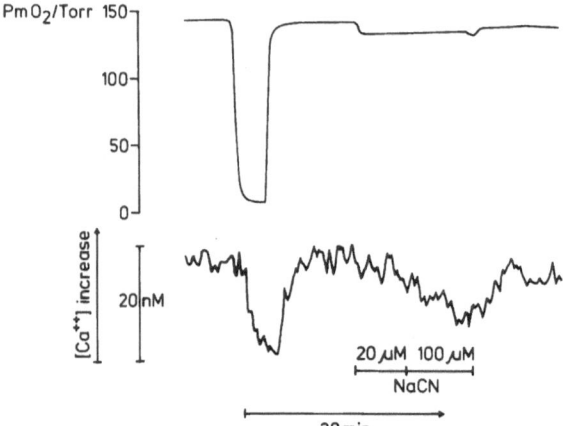

Fig. 2. Reaction of the intracellular calcium in endothelial cells to hypoxia and NaCN sensations in the superfusion medium. Calcium is measured fluorometrically with the calcium-sensitive fluorescence dye Indo 1. A calcium change of 20 nM is given by a *bar*. Cyanide was applied to the solutions in two different concentrations. The time of 30 min is indicated

2. To investigate this question in more detail, studies were carried out with cloned endothelial cells of rat brain microvessels, which show typical growth characteristics and produce factor VIII antigen [26]. Under the assumption that the intracellular calcium level acts as a second messenger for an as yet not identified PO_2 sensor, the calcium-sensitive fluorescence dye Indo 1 was used for intracellular calcium monitoring [7]. The ratio between the isosbestic point of the fluorescense spectrum at 445 nm and the fluorescence maximum of the spectrum at 480 nm was monitored continuously for relative changes. Figure 2 shows that superfusing endothelial cells grown on glass coverslips either with hypoxic saline solution or 1 mM NaCN results in a decrease of the intracellular calcium. If one considers the occurrence of contractile elements in endothelial cells, as described by Weigelt et al. [39], the decreased intracellular calcium content in endothelial cells under hypoxia could be interpreted as an expression of relaxation of contractile elements leading perhaps to the hypoxic vasodilatation in the carotid body. Since cyanide induces the same intracellular calcium changes, participation of the respiratory chain in the O_2 sensing process can be assumed.

Table 1. Dopamine and noradrenaline content in the cat carotid body under different O_2 conditions

	Normoxia	Hypoxia
Dopamine (pmol/glomus)	451 ± 200 (8)	315 ± 157 (8)
Noradrenaline (pmol/glomus)	617 ± 184 (8)	391 ± 177 (8)

3. As a further event carotid body secretory cells, called type-I cells, release transmitters under hypoxic situations, which generate action potentials in postsynaptic afferent nerve endings, a signal which regulates ventilation and blood circulation. Table 1 shows that the noradrenaline and dopamine content of the cat carotid body is significantly reduced under hypoxia [35].
4. The regulation of the aerobic glycolysis [17, 38] might represent a PO_2 sensing mechanism, which influences tumor growth. Aerobic glycolysis means the breakdown of glucose to lactate in the presence of oxygen, or in other words, in the presence of an active respiratory chain. Under normal oxygenation conditions the lactate formation is suppressed by the respiratory chain (Pasteur effect). However, some tissues like retina, renal medulla, or tumor tissue exhibit a missing Pasteur effect. In the case of tumor tissue it is reasonable to assume that pH and oxygen can influence tumor cell proliferation. Recent studies have indicated that DNA synthesis and cell proliferation are inhibited at intracellular pH values below 7.2 [18, 27] as well as the fact that nuclear DNA replication in Ehrlich ascites cells is PO_2 dependent [28]. Therefore, we have chosen tumor tissue in the form of multicellular spheroids [6]. Multicellular spheroids are aggregates of human and rodent tumor cells, which mimic the nodular tumor growth in tissue culture. Due to their three-dimensional growth, tumor cells in multicellular spheroids have comparable radio resistence and drug sensitivity with identical cells in vivo, which is often not the case when these cells are grown as monolayers. Figure 3 shows PO_2- and pH microelectrode measurements in a multicellular spheroid of human malign glioma cells. It can be seen that a PO_2 gradient as well as a pH gradient exist in the spheroid. The PO_2 gradient is created by oxygen consumption, whereas the pH gradient seems to be caused solely by glycolysis [8]. The quotient between the PO_2 gradient and the pH gradient is related to the volume doubling time of multicellular spheroids in the exponential growth phase. Human colon carcinoma spheroids have a doubling time of about 4 days and a quotient of about 500, whereas human malign glioma spheroids have a doubling time of about 8 days with a quotient of about 200 (Carlsson and Acker, unpublished observations). This means that fast-growing spheroids favor the oxygen consumption, whereas slowly-growing spheroids depend in their growth characteristics more on aerobic lactate production hinting to the importance of PO_2 and pH for growth regulation.

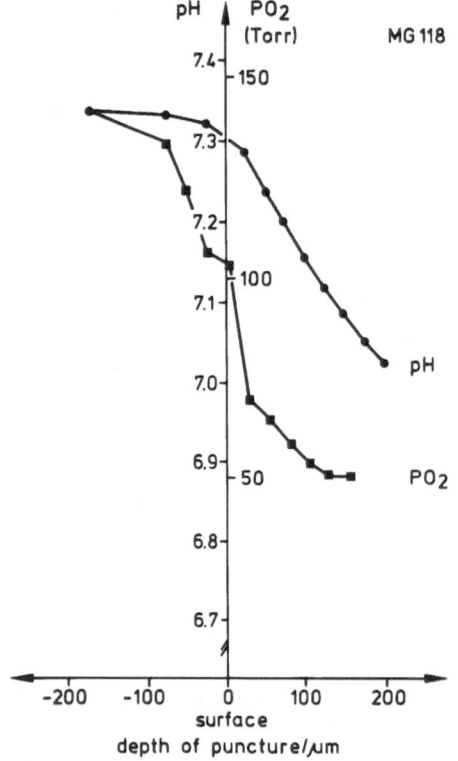

Fig. 3. PO_2 and pH gradients in a human malign glioma spheroid (MG118) measured with microelectrodes. The depth of puncture is given as positive values for the penetration inside the tissue towards the center and as negative values for measurements in the superfusion bath towards the surface

Basic Properties

For exploring the molecular mechanism of the oxygen sensing process in the above-introduced models (endothelial cells, carotid body, multicellular spheroids) different possibilities have to be considered.

1. Due to the direct interaction of oxygen with the respiratory chain, the regulation of ATP production by this process could be a signal for the cell to change activities. Since the critical mitochondrial PO_2 is far below 1 torr [36], steep PO_2 gradients in the cytosol are required to get the respiratory chain working as a PO_2 sensor under normal oxygen supply conditions. However, investigations by Wilson et al. [41] reveal that the critical PO_2 for the respiratory chain might be found at higher values, because cytochrome c could be reduced constantly in their experiments beginning with a PO_2 of 100 torr by interaction with the ATP/ADP quotient. Figure 4, shows the behavior of cytochrome c in a rat malign glioma spheroid, measured photometrically with a dual-wavelength procedure at the microscopical level [3], in relationship to a changed oxygen supply. Figure 4a gives the behavior of the tissue PO_2 at a depth of 200 μm measured inside the spheroid with a microelectrode in dependence on the PO_2 in the superfusion medium. The tissue PO_2 follows the medium PO_2 in a characteristic manner with a tight coupling during decrease, whereas increasing the medium

a

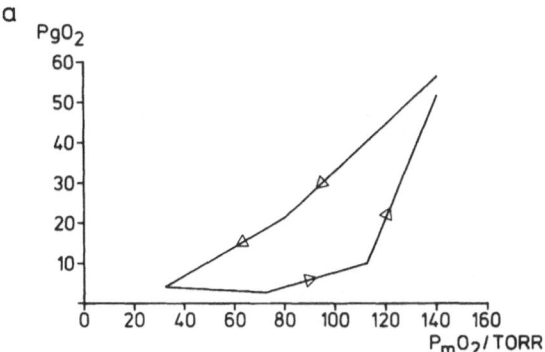

b

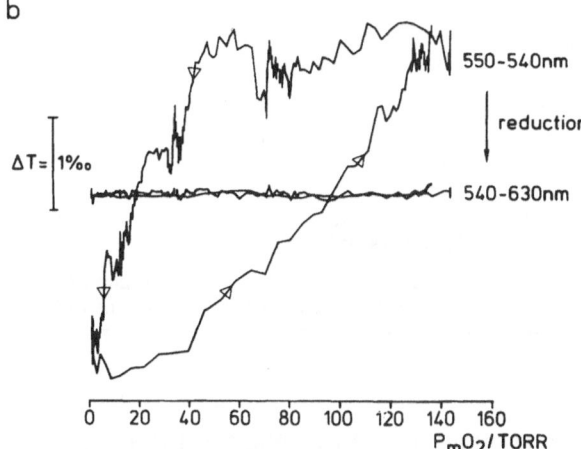

Fig. 4a, b. Dependence of tissue PO_2 (a) and cytochrome c signal (b) in a rat malign glioma spheroid on variations of the PO_2 in the superfusion bath. ◁ means lowering the PO_2, ▷ means increasing the PO_2. Cytochrome c was measured with the quotient of the light absorbance changes at 540 nm (isosbestic point) and 550 nm (absorbance maximum). The constancy of the isosbestic point was controlled by the quotient of the light absorbance changes at 540 and 620 nm

PO_2 is not promptly accompanied by an elevated tissue PO_2, but a higher medium PO_2 is necessary to induce an increase of the tissue PO_2. This hysteresis-like behavior, which can be explained by diffusion characteristics of oxygen, is also to be seen in the reduction and oxidation curve of cytochrome c. This means that cytochrome c follows exactly the PO_2 course indicating a lack of a critical PO_2 and confirming the findings of Wilson et al. [41], as mentioned above. It is not known at present, whether a steep cytosolic PO_2 gradient can account for the discrepancy in the critical mitochondrial PO_2 between isolated mitochondria and mitochondria in intact tissue. However, these results demonstrate the important function of the respiratory chain in the oxygen sensing process.

2. As a further candidate the lactate dehydrogenase (LDH) with its different isoenzymes seems to be appropriate. LDH exists in five different isoenzymes, which are tetrameres of two subunits, H, and, M. The synthesis of the different forms of the enzyme from LDH_1 to LDH_5 can be influenced by oxygen changes. In lymphocytes and chicken embryos hypoxia induces the formation of the M types of the isoenzymes with a higher affinity for pyruvate [21]. The time constant for these changes seems to be short, since 10 min of hypoxia are sufficient

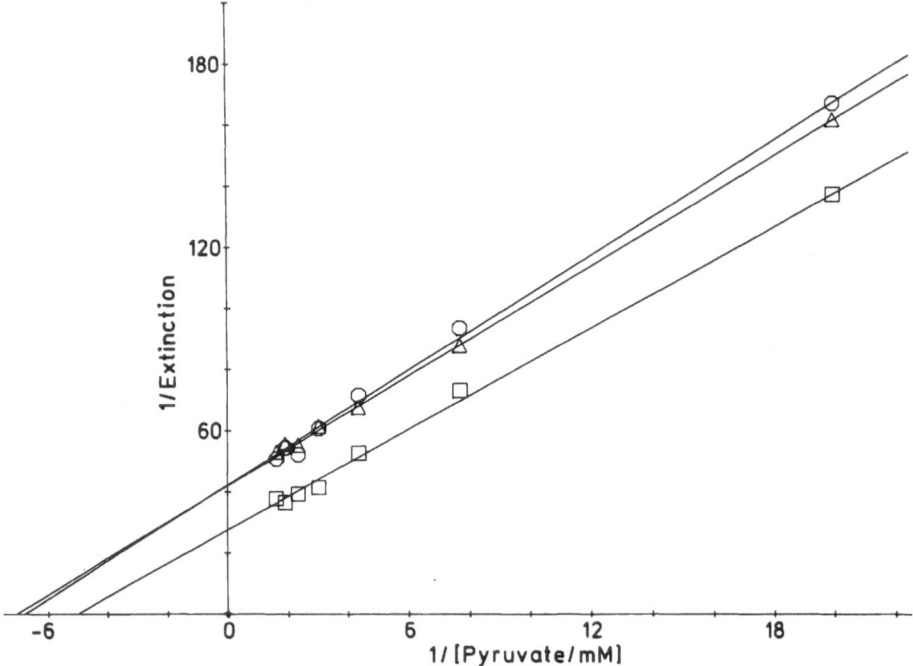

Fig. 5. Lineweaver-Burk plot of the LDH Michaelis-Menten kinetic in rat malign glioma spheroids. The *upper two lines* are the values for air exposure, the *lower line* for hypoxia. The turnover
velocity of pyruvate by LDH is given in arbitrary units, as measured in the photometer

to induce a change in the maximal pyruvate turnover (V_{max}) of LDH, as shown
in Fig. 5. This figure demonstrates the Lineweaver-Burk plot of the Michaelis-
Menten kinetic for LDH, which was isolated from rat malign glioma tumor cells
grown as multicellular spheroids. Whereas two spheroids were kept under air
conditions, one other was exposed to hypoxia. It can be seen that these short
exposures induced a significant increase in V_{max} with decreasing the affinity for
pyruvate. This finding might be interpreted that hypoxia has an influence on the
allosteric conformation of LDH. Chronic or acute changes in the oxygen supply
are accompanied by changes of the LDH activity, which then influence the respiratory chain, because both compete for NADH,H^+. This competition can be
demonstrated by measurements of oxygen consumption under oxamic acid conditions. Oxamic acid competes with pyruvate for LDH binding and increases
the oxygen consumption in Ehrlich ascites tumor cells with a concomitant depression of the lactate formation [24]. Table 2 shows the behavior of the oxygen
consumption of rat malign glioma tumor cells grown as multicellular spheroids
under oxamic acid, which reveals under these conditions a nearly 100% increase
vs. control. If one considers the pivotal role of PO_2 and pH in the proliferation
rate of tumor cells, the competition between LDH activity and the respiratory
chain for NADH,H^+ seems to be decisive, especially under the point of view
that some tumor types, for instance, human brain tumors typically reveal an M

Table 2. Influence of oxamate on the oxygen consumption of multicellular spheroids

	Oxygen consumption (μl O$_2$/min/spheroid)
Glucose 1 g/l	10.2 ± 0.85 (7)
Glucose 1 g/l + 40 mM Na-oxamate	19.5 ± 1.32 (7)

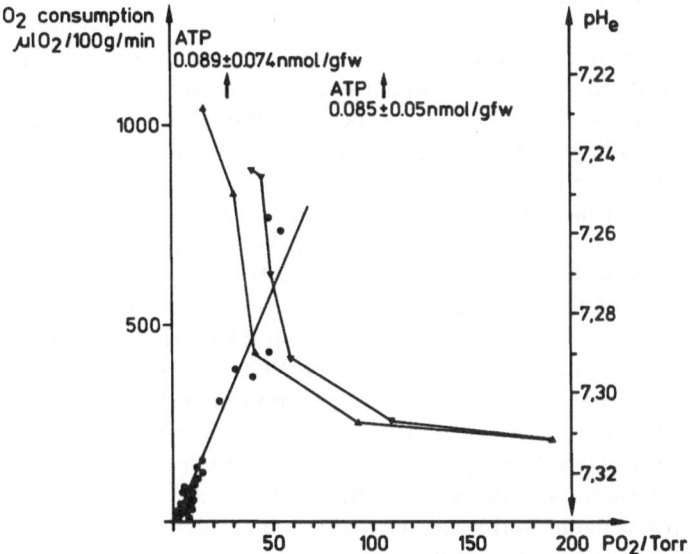

Fig. 6. Dependence of oxygen consumption and extracellular tissue pH in the cat carotid body on the prevailing PO$_2$. Furthermore, the ATP content of the organ is shown under normoxic and hypoxic conditions

type LDH isoenzyme [37]. This competition also seems to exist in the carotid body as an important step to liberate transmiters from type-I cells during hypoxia for exciting nerve endings. The respiratory chain of the carotid body mitochondria has often been discussed to be the most probable candidate for a PO$_2$ sensor. Anichkow and Belinki [9] as well as Joels and Neil [15] assumed that chemosensory excitation, especially under hypoxia, is caused by a decrease in ATP levels in the carotid body tissue. Biscoe [10] proposed that energy depletion under hypoxia triggers the nervous discharge by producing membrane instability in sensory nerve endings. Using several inhibitors and uncouplers of the respiratory chain, Mulligan et al. [22] gave further support to the idea that oxidative phosphorylation is involved in PO$_2$-, but not in PcO$_2$ chemoreception of the carotid body. The mitochondria of the carotid body seem to be specialized, since they have a PO$_2$-dependent oxygen consumption [8, 20, 29, 40], as demonstrated in Fig. 6. The absolute level of the oxygen consumption depends on the method which is applied to measure this parameter. Using Fick's principle for determining the oxygen consumption [29], i.e., arteriovenous oxygen

concentration difference times flow, reveals values of about 9 ml $O_2/100$ g/min, whereas methods like tissue PO_2 disappearance curves in the carotid body after perfusion stop [4, 40], or applying a modified Warburg principle on the excised carotid body [20] result in values between 1–2 ml $O_2/100$ g/min. At the moment, this discrepancy stands unexplained in the literature. Further experiments might investigate, whether carotid body mitochondria are able to produce a short circuit for protons over the inner mitochondrial membrane, thereby increasing the oxygen consumption and perhaps meaning that the carotid body perfusing blood contains substances, which are able to act as protonionophores [23]. As can be seen in Fig. 6, the PO_2-dependent oxygen consumption is accompanied by a change in tissue pH. Concomitantly with decreasing oxygen consumption under hypoxia, an acidification can be measured [12], which favors the participation of aerobic glycolysis on the chemoreception. This means that ATP production in the glycolytic pathway is increased and the net production of H^+ ions by ATP hydrolysis cannot be compensated any more by the respiratory chain [11]. The PO_2-dependent oxygen consumption could, therefore, be caused by the competition for $NADH,H^+$ between LDH and the respiratory chain. The participation of LDH is in agreement with findings of Petrova [25] showing that 20 min of hypoxia induced in the rat carotid body a distinct increase of an LDH isoenzyme mostly composed of M subunits. Respiratory chain and glycolysis together are able to maintain a constant ATP level in the carotid body under hypoxia, as described by Acker and Starlinger [5]. One might, therefore, assume that a decresed oxygen consumption under hypoxia also means a decreased membrane potential of the carotid body mitochondria, leading to an impaired calcium buffer capacity [23] with an increased cytosolic calcium level and a consequent transmitter release. In general, the importance of the cytosolic calcium level for transmitter release in secretory cells is very well known [32]. This was also shown for the carotid body by Grönblad et al. [13]. By using electron microscopy they could demonstrate an increased exocytosis of transmitter-containing vesicles of type-I cells after external application of ionophore A23187, which is known to increase the cytosolic calcium activity.
3. Oxidases and oxygenases with K_mO_2 values between 30 and 300 torr could also be involved in the O_2 sensing process. About 28 different types are known [16]. In the case of the carotid body the involvement of an oxidase seems to be probable, which was described previously in skunk cabbage and mung bean [31]. In the mouse carotid body Acker and Eyzaguirre [3] were able to show that hypoxia and cyanide give different absorption spectra between 500 nm and 620 nm. The cyanide spectrum could be used to calculate the cytochrome content of the carotid body with $1.89 \cdot 10^{-3}$ μM/g fresh weight for cytochrome aa$_3$. These calculated concentrations are close to values reported for the white musculature of the rabbit [30]. If we assume that the cyanide spectrum is a genuine expression of the maximal absorption of different cytochromes, a quotient cytochrome c/cytochrome aa$_3$ of 1.9 is close to values reported for the kidney and suggests a normal composition of both cytochromes. The hypoxic sensitive light absorbance change with a light transmittance increase peaking at 540 nm might represent a cyanide-insensitive part of the respiratory chain in the carotid body, probably comparable with the plant oxidase, which can be inhibited by hydroxamic

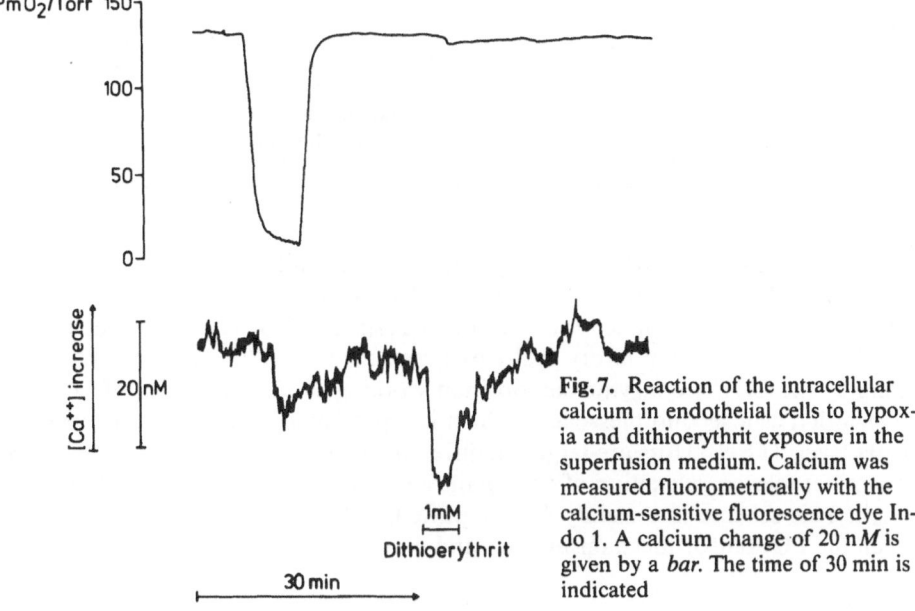

Fig. 7. Reaction of the intracellular calcium in endothelial cells to hypoxia and dithioerythrit exposure in the superfusion medium. Calcium was measured fluorometrically with the calcium-sensitive fluorescence dye Indo 1. A calcium change of 20 nM is given by a *bar*. The time of 30 min is indicated

acids [31]. This oxidase might act as the primary PO_2 sensor in the carotid body and to install a hypothesis switch on the aerobic glycolysis with decreasing PO_2. Figure 7 demonstrates results which might hint to the involvement of another oxidase in the PO_2 sensor mechanism. As already shown in Fig. 2, cloned endothelial cells grown on glass coverslips were superfused with saline solution. Lowering the PO_2 in the superfusion medium results in a decrease of the intracellular calcium, as measured with the intracellular calcium-sensitive fluorescense dye Indo 1 [7]. The same change could be induced by dithioerythrite, an SH group donator. This finding might imply the possible involvement of glutathione peroxidase in the PO_2 sensor mechanism. Glutathione peroxidase reduces peroxides with concomitant oxidation of reduced glutathione (GSH) to form the corresponding thiol and glutathione disulfate (GSSG) [19]. The transition in the thiol redox state has profound effects on active and passive cation permeability as well as on the chemical stimulation of excitable cells [34]. Therefore, it is thinkable, that a Ca^{2+} ATPase is activated by this mechanism in the endothelial cells under hypoxic conditions leading to a cytosolic calcium clearance with a consequent vasodilatation in microvessels. The activation of a Ca^{2+} ATPase, a thiol-rich protein, by reducing agents splitting disulfide bonds was also shown in red cells [14].

Conclusions

Table 3 shows an overview of the different PO_2 sensor mechanisms, as described in this article, and their arrangement in the whole PO_2 sensor process.

Table 3. Schematic pathway of the PO_2 sensor process

Stimulus	Sensor	Effector	Effect
PO_2	Respiratory chain	Ca^{2+}	Transmitter release
	LDH isoenzymes	pH	Endothelial cell relaxation
	Oxidases		Tumor cell proliferation
	GSSG/GSH		

The PO_2 can be regarded as the specific stimulus and is determined in the tissue by diffusion constants, vasculature, local flow, oxygen consumption, arterial PO_2, and O_2 transport capacity (for review see [2]). The PO_2 is then interacting inside the cell with the respiratory chain, oxidases, the GSSG/GSH redox system, or changing the LDH isoenzyme composition working as a sensor. The effector can be regarded as a second messenger, which is represented by the cytosolic calcium or pH level. The effectors are targeting different cell elements in the cell leading to transmitter release, relaxation of contractile elements, or controlling of DNA replication. This scheme is certainly too simplified, but might help to distinguish between the PO_2 sensor mechanism itself and secondary events.

References

1 Acker H (1980) The meaning of tissue PO_2 and local blood flow for the chemoreceptive process of the carotid body. Fed Proc 39: 2641–2647
2 Acker H (1983) Tissue oxygen transport in health and disease. In: Pallot DJ (ed) Control of respiration. Croom Helm, London, pp 157–202
3 Acker H, Eyzaguirre C (1987) Light absorbance changes in the mouse carotid body during hypoxia and cyanide poisoning. Brain Res 409: 380–385
4 Acker H, Lübbers DW (1977) The kinetic of local tissue PO_2 decrease after perfusion stop within the carotid body of the cat in vivo and in vitro. Pflügers Arch 369: 135–140
5 Acker H, Starlinger H (1984) Adenosine triphosphate content in the cat carotid body under different arterial O_2 and CO_2 conditions. Neurosci Lett 50: 175–179
6 Acker H, Carlsson J, Durand R, Sutherland RM (eds) (1984) Spheroids in cancer research. Recent results in cancer research, vol 95. Springer, Berlin Heidelberg New York
7 Acker H, Pietruschka F, Dufau E (1987a) The effect of hypoxia and cyanide on intracellular calcium in cloned endothelial cells of brain microvessels. Pflügers Arch 408: R72
8 Acker H, Carlsson J, Holtermann G, Nedermann Th, Nylen T (1987b) Influence of glucose and buffer capacity in the culture medium on growth and pH in spheroids of human thyroid carcinoma and human glioma origin. Cancer Res 47: 3504–3508
9 Anichkov SK, Belinki MR (eds) (1963) Pharmacology of the carotid body chemoreceptors. Pergamon, Oxford
10 Biscoe T (1971) Carotid body: structure and function. Physiol Rev 51: 437–495
11 Busa WB, Nuccitelli R (1984) Metabolic regulation via intracellular pH. Am J Physiol 246: R409–R438
12 Delpiano MA, Acker H (1985) Extracellular pH changes in the superfused cat carotid body during hypoxia and hypercapnia. Brain Res 342: 273–280
13 Grönblad M, Akerman KE, Eränko O (1979) Induction of exocytosis from glomus cells by incubation of the carotid body of the rat with calcium and ionophore A23187. Anat Rec 195: 387–395
14 Hebbel RP, Shalev O, Foker W, Rank BH (1986) Inhibition of erythrocyte Ca^{2+} ATPase by activated oxygen through thiol- and lipid-dependent mechanisms. Biochim Biophys Acta 862: 8–16
15 Joels M, Neil E (1963) The excitation mechanism of the carotid body. Br Med Bull 19: 21–24

16 Jones DP (1986) Renal metabolism during normoxia, hypoxia and ischemic injury. Annu Rev Physiol 48: 33–50
17 Krebs HA (1972) The Pasteur effect in the relation between respiration and fermentation. Essays Biochem 8: 1–34
18 L'Allemain GL, Franchi A, Cragoe E, Pouysségur J (1984) Blockade of the Na^+/H^+ antiport abolishes growth factor induced DNA synthesis in fibroblasts. J Biol Chem 259: 4314–4319
19 Lash HL, Jones DP, Orrenius ST (1984) The renal thiol (glutathione) oxidase subcellular localization and properties. Biochim Biophys Acta 779: 191–200
20 Leitner LM, Liaubet MJ (1971) Carotid body oxygen consumption of the cat in vitro. Pflügers Arch 323: 315–322
21 Lindy S, Rajasalin M (1966) Lactate dehydrogenase isoenzymes of chicken embryo: response to variations of ambient oxygen tension. Science 153: 1401–1403
22 Mulligen E, Lahiri S, Storey BT (1981) Carotid body O_2 chemoreception and mitochondrial oxidative phosphorylation. J Appl Physiol 51: 438–446
23 Nicholls DG (1982) Bioenergetics: an introduction to the chemiosmotic theory. Academic, London
24 Papaconstantinou J, Colowick SP (1961) The role of glycolysis in the growth of tumour cells I. Effects of oxamine acids on the metabolism of Ehrlich ascites tumor cells in vitro. J Biol Chem 236: 278–284
25 Petrova NU (1974) Effect of hypoxia on the lactate dehydrogenase isoenzyme composition in the rat carotid body. Bull Exp Biol Med 78: 1005–1006
26 Pietruschka F, Acker H (1986) Production of angiogenetic factors by tumour cells growing in spheroid or monolayer cultures. Eur J Cell Biol 42: 48
27 Pouysségur J (1985) The growth factor activatable Na^+/H^+ exchange system: a genetic approach. Trends Biochem Sci 2: 453–455
28 Probst H, Gekeler V, Helftenbein E (1984) Oxygen dependence of nuclear DNA replication in Ehrlich ascites cells. Exp Cell Res 154: 327–341
29 Purves MJ (1970) The effect of hypoxia, hypercapnia and hypotension upon carotid body blood flow and oxygen consumption in the cat. J Physiol (Lond) 209: 395–416
30 Schollenmeyer P, Klingenberg M (1962) Über den Cytochromgehalt tierischer Gewebe. Biochem Z 335: 426–439
31 Schonbaum GR, Bonner WD, Storey PT, Bahr JT (1971) Specific inhibition of the cyanide-insensitive respiratory pathway in plant mitochondria by hydoxamic acids. Plant Physiol (Bethesda) 42: 124–128
32 Schulz J, Stolze HH (1980) The exocine pancreas. The role of secretagogues cyclic nucleotides and calcium in enzyme secretion. Annu Rev Physiol 42: 127–156
33 Seidl E, Acker H, Teckhaus L (1979) Quantitative Erfassung des Gefäßvolumens des Glomus caroticum der Katze unter den Bedingungen der Normoxie, Hypoxie und Hypercapnie. Microsc Acta 3: 185–189
34 Sies H (1977) Peroxisomal enzymes and oxygen metabolism in liver. In: Reivich M, Coburn R, Lahiri S, Chance B (eds) Tissue hypoxia and ischemia. Plenum, New York, pp 51–65
35 Starlinger H, Acker H (1986) The norepinephrine and dopamine content of the cat carotid body in vivo under normoxic an hypoxic conditions. Neurosci Lett 64: 65–68
36 Starlinger H, Lübbers DW (1973) Polarographic measurements of the oxygen pressure performed simultaneously with optical measurements of the redox state of the respiratory chain in suspension of mitochondria under steady state conditions at low oxygen tension. Pflügers Arch 341: 15–22
37 Vivekanadan S, Ramalakara Rao AP, Schwam R, Ranaka TS (1982) Sequential determination of cerebrospinal fluid lactate dehydrogenase isoenzyme in human brain tumours as treatment. Acta Neurol Scand 66: 347–354
38 Warburg OH (1962) New methods of cell physiology. Thieme, Stuttgart
39 Weigelt H, Fujii F, Lübbers DW, Hauck G (1981) Specialized endothelial cell in frog mesentery. Attempt of an electrophysiological characterization. Bibl Anat 20: 89–93
40 Whalen WJ, Nair P, Sidebotham T, Spander J, Lacerna M (1981) Cat carotid body. Oxygen consumption and other parameters. J Appl Physiol 50: 129–133
41 Wilson DF, Owen CS, Erecinska M (1979) Quantitative dependence of mitochondrial oxidative phosphorylation on oxygen concentration: a mathematical model. Arch Biochem 195: 494–504

II. Cell Physiology

Oxygen Dependent Regulation of DNA Replication of Ehrlich Ascites Cells In Vitro and In Vivo*

H. Probst and V. Gekeler

Physiologisch-chemisches Institut der Universität Tübingen, 7400 Tübingen 1, FRG

Introduction

Control of cell growth in higher animals normally occurs by switching between a resting and a cycling state. In most cases resting occurs at 2 c DNA content (G_0 phase). The metabolism of resting cells is organized principally different to that of cycling cells, which generally are committed to DNA synthesis. When resting cells receive a switching signal, they usually require several hours before they can enter DNA synthesis. It is commonly believed that DNA synthesis, once started, proceeds relatively autonomously following a more or less fixed S-phase program [1] which typically runs for 5-8 h. This program temporally and spatially organizes the activation of groups (clusters) of replication units (replicons) [1]. Different sets of clusters are activated at different times in the S phase. The sets are thought to comprise functionally related genes [2]. In cycling cells, the synthesis of DNA and of DNA-associated protein occupies a considerable portion of total metabolic activity. This concerns both the consumption of organic matter as well as of ATP energy. The present paper deals with an oxygen-dependent regulation system acting by selectively suppressing or re-triggering activation of replicon clusters in Ehrlich ascites cells. The regulation sets in before hypoxia imposes measurable metabolic constraints to the cells.

We suggest that this might represent a surviving principle for early animal embryos. Cells of early embryos are usually still lacking an O_2 transporting blood circulation and, in the case of mammals, are not yet connected to the maternal circulation. Therefore, hypoxic conditions are likely to occur frequently in the microenvironment of these cells. In this situation, a mechanism is of advantage that allows to temporarily switch off or greatly diminish ongoing DNA synthesis before its additional requirements contribute to a generalized metabolic collapse. However, simply halting active replication forks would release parts of the genetic information in a replicative state and thus render them inaccessible when required for maintaining basic vital functions. Control at the hierarchic level of replicon clusters, as observed in the present case, appears to be distinctly more suitable. In the case of tumor cells, a regulation of this kind will improve their adaptability to

---

* Dedicated to Prof. Dr. Fr. Schneider on the occasion of his 60th birthday

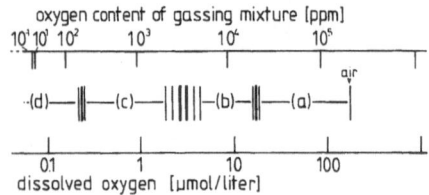

Fig. 1. Ranges of hypoxic stringency (from [5])

the supply situation and thus contribute to their malignant potential. Earlier reports, discussed in detail in [3-5], demonstrated that hypoxia affects cell-cycle progression and net DNA synthesis of animal cells. Our results, surveyed in this paper, provide means for understanding these observations at the control level of the course of genome replication in higher cells. They imply consequences for modeling of tumor growth and for designing therapeutic regimens.

Studies on Cultured Cells

Stringency Ranges of Hypoxia

We gassed all cultures with gas mixtures (argon/5% CO_2) containing different portions of O_2. Cell growth, cell cycle progression, and different aspects of DNA replication were examined during gassing and after reestablishment of "aerated" conditions [5]. The results prompted us to define four hypoxic stringency ranges: (a) range of no response; (b) range of inconstant and/or incomplete response; (c) range of reversible shut down of DNA synthesis; and (d) range of hypoxic cell damage.

Figure 1 displays the situation of these ranges relative to the O_2 content of the gassing mixtures used, and relative to the concentration of dissolved O_2 in the culture fluid, measured between 8 and 12 h after start of gassing. At that time an equilibrium or, at very low O_2 contents of the gassing mixture, a "quasi-steady-state" had been attained [5].

DNA Replication

The DNA synthesis rate in cultures gassed with mixtures containing O_2 portions within range (c) typically declined in a decay-like course. Upon reoxygenation, a quick recovery occurred, which exceeded the DNA synthesis rate of the aerobic control culture the more, the longer the preceding hypoxic period lasted (Fig. 2C). In the range of hypoxic cell damage (d), below 150 ppm O_2, the decline was more pronounced. However, the recovery was progressively abolished when the hypoxic period was prolonged or hypoxia further strengthened (Fig. 2A, B).

The length distributions of growing daughter DNA strands reflect the relative frequency of replicon initiations. A decrease will diminish the proportion of recently initiated short chains, and a reincrease will reelevate them [4]. The length distribution can be evaluated by alkaline sedimentation analysis of DNA labeled by [³H]dThd pulses of 5-10 min. Using a pulse/chase labeling schedule, alkaline sedimentation analyses is also suited to demonstrate the maturation to the large

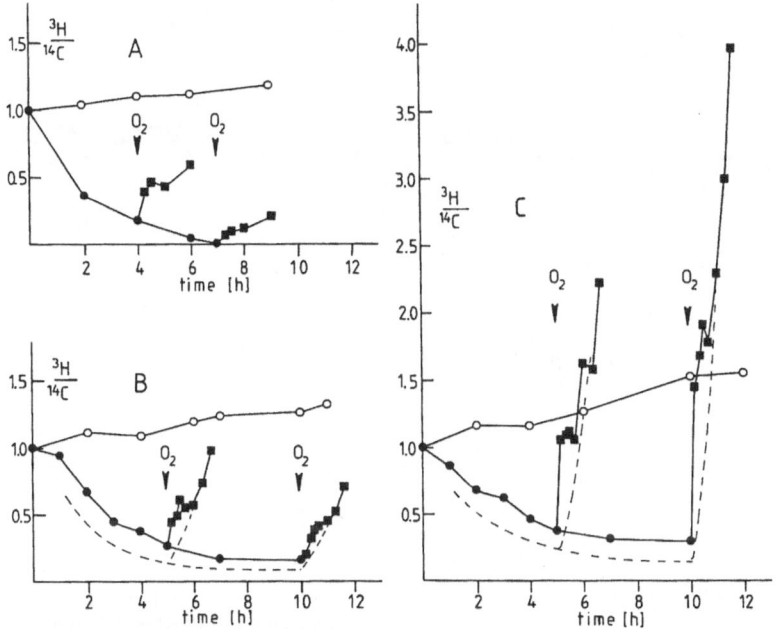

Fig. 2 A–C. DNA synthesis rate during hypoxic gassing and after reoxygenation. Cell cultures prelabeled with ^{14}C-thymidine were divided 2 h after transfer to fresh medium and one-half was gassed. Parts were reoxygenated at the times indicated by *arrows*. Samples were incubated under aerobic or hypoxic conditions, respectively with [^{3}H]dThd for 8 min and then processed for determination of the acid insoluble radioactivity. **A,** gassing with <0.1 ppm O_2; **B,** 100 ppm O_2; **C,** 200 ppm O_2; ○, aerobic control; ●, hypoxic; ■, reoxygenated. The results are expressed in terms of ^{3}H/^{14}C quotients which were reduced to the first quotient of each curve = 1 (from [5])

continuous DNA strands of daughter chromatids. Figure 3 displays a series of sedimentation profiles recorded in the course of a deoxygenation/reoxygenation experiment [3]. The progressive depletion of short pulse labeled DNA chains during the hypoxic period and their reappearance after oxygen recovery is clearly visible. The latter is not observed when the gassing mixture contains <150 ppm O_2. Figure 4 shows the profiles obtained before and after oxygen readmittance at the ends of the experiments of Fig. 2 A and C. The absence of short recently initiated chains after reoxygenation following gassing with the severely hypoxic gas mixture (Fig. 4 A; 0.1 ppm O_2) distinctly contrasts with the results obtained after gassing with 200 ppm O_2 (Fig. 4 B). Obviously, the severely hypoxic cells lost the ability to make up for replicon initiation omitted during the O_2 deficiency. This is consistent with the poor and slow recovery of the overall DNA synthesis rate (Fig. 2 A). However, the profiles obtained after reoxygenation from 200 ppm O_2 revealed a burst of initiations (Fig. 4 B). This burst easily explains the excess of the DNA synthesis rate observed at the same time (Fig. 2 C). The increasing tendency of this excess with increasing duration of hypoxia suggests that missing initiations were progressively accumulated, ready to be immediately activated when oxygen was readmitted.

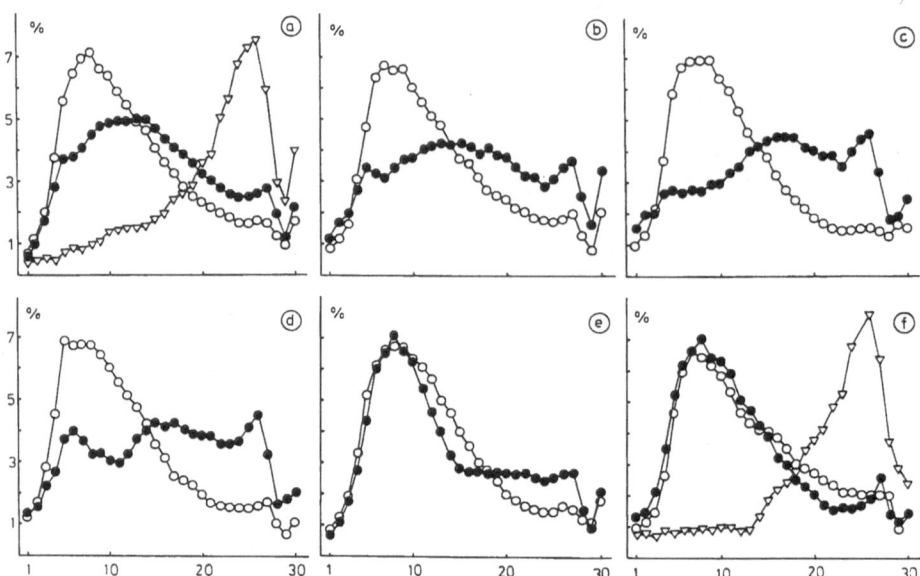

Fig. 3 a–f. Alkaline sedimentation profiles of DNA of cells pulse labeled by [³H]dThd under hypoxic conditions (**a**, 1.5 h; **b**, 3 h; **c**, 5 h) and after reoxygenation following 5 h hypoxia (**d**, 20 min; **e**, 1 h; **f**, 3 h). A culture of cells prelabeled with ¹⁴C-thymidine was divided. One part was incubated aerobically (control), the other was deoxygenated by gassing with argon/CO_2 mixture containing about 200 ppm O_2. The labeled cell samples of the control culture (−O−O−) and of the deoxygenated/reoxygenated culture (−●−●−) were lysed on the top of alkaline SW 28 Ti gradients and centrifuged for 6 h at 26000 rpm and 23° C. *Ordinate,* percent of total counts; *abscissa,* gradient fraction number. One fraction corresponds to 5.4 S. For the sake of clarity, only two of the 12 (nearly indentical) ¹⁴C-profiles (−▽−▽−), representing maturated DNA, are shown (from [3])

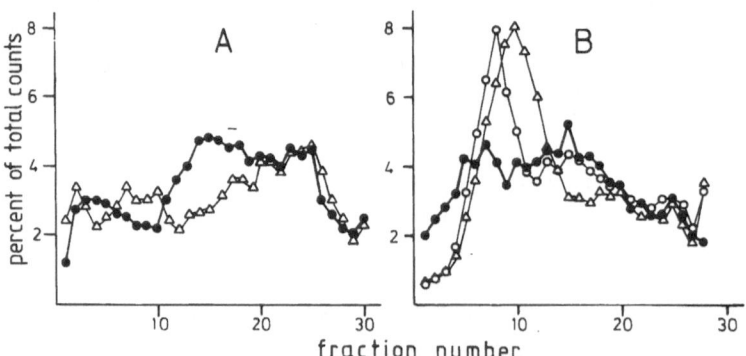

Fig. 4 A, B. Alkaline sedimentation profiles of DNA of cells pulse labeled under hypoxia (−●−●−) after 7 h gassing with 0.1 ppm O_2 (**A**) and 10 h gassing with 200 ppm O_2 (**B**) and 20 min (−O−O−) and/or 40 min after reoxygenation (−△−△−) (from [5])

Table 1. Results of DNA fibre autoradiography

	I	II	III	IV	V
Relative initiation frequency	1.36	0.164	2.66	2.15	2.12
Sample size	340	156	242	246	215
Fork progress rate					
(Mean ± SD (μm/min))	1.15 ± 0.63	0.77 ± 0.41	0.86 ± 0.43	0.80 ± 0.43	1.19 ± 0.65
Sample size	263	201	128	152	110
Replicon clustering					
Replicons per cluster (mean ± SD)	2.8 ± 1.0	2.9 ± 1.1	2.9 ± 1.0	3.0 ± 1.4	2.9 ± 1.0
Cluster length[a] (μm) (mean ± SD)	188 ± 118	228 ± 139	203 ± 111	162 ± 118	231 ± 121
Sample size	62	49	59	52	57
Inter initiation distance (μm)					
Mean ± SD	94.0 ± 49.8	104.0 ± 56.9	86.3 ± 51.0	72.5 ± 38.7	110.3 ± 51.3
Median	80.6	90.7	72.3	61.8	96.6
Sample size	144	51	114	93	104
Gap length (μm)					
Mean ± SD	57.8 ± 59.4	59.4 ± 48.6	77.3 ± 77.7	75.3 ± 85.9	49.1 ± 31.6
Sample size	97	79	41	49	52

A freshly established cell culture was grown in air for 2 h and then divided into five samples (I–V). Sample I (control) was further incubated under air for 2 h and then labelled for DNA fibre autoradiography. Sample II was labelled beginning at 2 h after O_2 removal (under hypoxic conditions). Samples III–V were reaerated after 5 h hypoxia. Labelling of sample III was started 10 min before reaeration; the labelling start of sample IV coincided with the reaeration and that of sample V was done 1 h after reaeration.
[a] Distance between the limits of the outmost tracks of "hot pulses" on both ends of a cluster.
From [4]

Above about 5×10^3 ppm O_2 in the gassing mixture, in range (b), principally the same effects could be observed as in range (c). However, the reactions of the cells were often inconstant or incomplete. Above 2×10^4 ppm, in range (a), replication was completely normal and indistinguishable from that observed under air.

The action of replicons in eukaryotic cells can be visualized by DNA fibre autoradiography [6]. The results obtained in a deoxygenation/reoxygenation experiment with about 200 ppm O_2 in the hypoxic gassing mixture were compiled in Table 1. The quotient of postpulse and prepulse initiation patterns, indicating the frequency of initiations during the "hot pulse" relative to preexisting active replicons (designated "relative initiation frequency"), distinctly decreased under hypoxia and, immediately after reoxygenation, increased in excess. This confirms the results obtained by alkaline sedimentation (Fig. 4 B). The decrease of the fork progress rate by about ⅓ does not necessarily reflect retardation of active forks. Rather, it might be a consequence of a depletion of short-lived replicons preferentially bearing the fast forks [4]. The coordiantion of replicon initiations in clusters and the spatial distribution of the initiation points within the clusters, obviously were not altered, neither by deoxygenation nor by reoxygenation.

It is possible to estimate the approximate "age" of a replicon which produced a prepulse initiation pattern from its fork movement rates and the distance of its

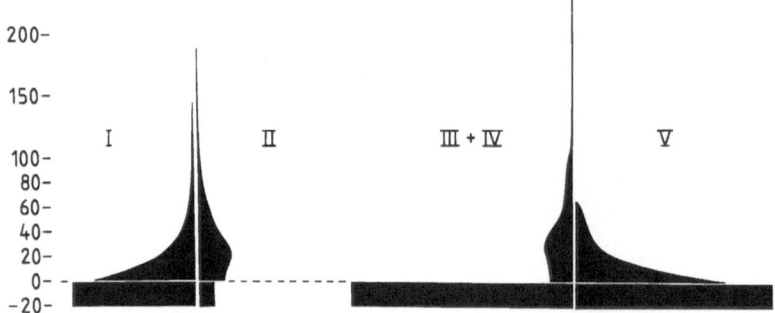

Fig.5. "Extended age pyramids" of the operating replicons evaluated by DNA fibre autoradiography in the experiment of Table 1. *Ordinate,* "replicon age" at the beginning of the "hot" pulse (min); *abscissa,* relative frequency of replicons (arbitrary units). The *rectangular areas* ("extension") represent the number of postpulse figures relative to the number of the prepulse figures forming the respective pyramid above. The areas of all four pyramids are equal (from [4])

forks ("gap" in [4]). We evaluated the prepulse patterns obtained from the five different cell populations labeled in the experiment of Table 1, and compiled the corresponding age distributions which are presented in the form of "age pyramids" in Fig. 5. The reference time of these distributions is the start of the hot pulse. Postpulse patterns allow no estimation of replicon age. However, the ratio of postpulse to prepulse patterns ("relative initiation frequency") indicates the relative number of replicons initiated within the duration of the hot pulse (20 min). This allows to extend the age pyramids from the reference time 20 min into the future, of course, without differentiation of individual ages.

The extended age pyramid of replicons active in the cell sample I (aerated control) clearly exhibits features of an expanding population. It possesses a broad basis and a slim apex. Its extension indicates further augmentation. The extended pyramid of sample II, the cells of which were labeled 2 h after deoxygenation start, exhibit features of a population distinctly dying out. When the hot pulse was started shortly before or coincidently with the reoxygenation after a 5-h hypoxic period (III and IV), the prepulse patterns obtained represent exclusively replicons initiated during hypoxia. Consistently, according to the pyramid alone, this population also seems to be dying out. However, abundant postpulse patterns, originating from an initiation burst during the hot pulse, prominently enlarged the area of the extension. One hour later (V), we found a picture representing a population still distinctly more expanding than the control. Note that the evaluated sample contained no replicon "older" than 60 min.

Alkaline sedimentation analysis of pulse/pulse-chase labeled cells [4] indicated that maturation of newly polymerized DNA to long chains was not significantly affected by hypoxia.

Growth and Cell-Cycle Progression

Parallel with the above experiments on DNA replication we recorded the growth of the gassed cultures and examined their cell-cycle progression by DNA flow cy-

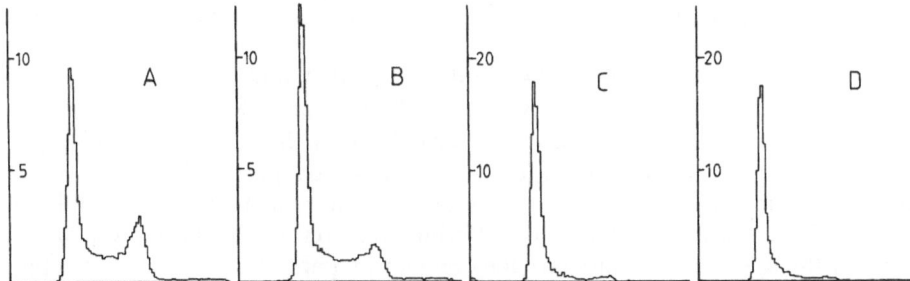

Fig. 6 A–D. Influence of gassing with 200 ppm O_2 on the DNA histograms of asynchronous cells and of G_1 cells selected by zonal centrifugation. The cultures **A** (aerated control) and **B** (gassed) were parts of the same starting culture and analysed after 12 h. **C** represents the G_1 cells directly after selection and **D** after 12 h hypoxic gassing (from [5])

tofluorometry and autoradiographic techniques [5]. In general, the results were in accordance with earlier work of others cited in [5].

When dissolved O_2 decreased below 3–6 μM, growth ceased, G_1 cells stopped to enter DNA synthesis, and the increase of DNA in S cells ceased or became distinctly retarded [5]. The relative portion of cells exhibiting G_1 DNA increased significantly. Figure 6 illustrates these statements by means of DNA histograms of asynchronous cells and of G_1 cells selected by a zonal centrifugation procedure developped by us [7].

All described effects were quickly reversible upon reoxygenation if the hypoxia remained above range (d): Growth was resumed, S cells continued in intensive DNA synthesis and, important to note, within a few minutes a large portion of the accumulated G_1 cells began to synthesize DNA.

Tentative Interpretation of the In Vitro Results

It appears reasonable to attribute the changes of cell growth and of cell-cycle progression to the changes of DNA replication occurring at the same O_2 conditions. They emerge when the concentration of dissolved O_2 decreases below 3–6 μM or 1.7%–3.5% of the atmospheric pO_2. This is quite a bit more than one order of magnitude above the O_2 tension reducing the respiration of ascites cells to ½ of the normal rate [8]. Thereby, the supply of intracellular ATP energy is virtually indistinguishable from that of normal aerated cells [5]. Also, no shortages of building blocks, neither for DNA nor for other cellular constituents, could ever be detected [9]. Consequently, the DNA polymerization events at active replication forks are found to continue virtually normally, and already initiated replicons are normally completed. Merely the initiation of new replication units, forthcoming according to the cellular program of sequential genome replication [1, 2], is suppressed. The suppression is coordinated at the hierarchic level of replicon clusters, where the control of eukaryotic replication is thought to occur [1]. Upon recovery of a higher oxygen supply, the burst-like return of initiations is coordinated at the same level. Moreover, it was reported [10] that the progress of the cells through the G_1 phase was not affected by hypoxia and that cells of the very late S phase attained full G_2

DNA content; cells which were in $G_2 + M$ at the beginning of hypoxia divided and entered G_1.

Considering that, we interpret the shutdown of DNA replication in the hypoxic range (c) as a regulatory event actively controlled by the cells. This regulation allows the cells to escape temporarily and orderly from the expensive task of genome replication or to diminish greatly its activity. By postponing the start of scheduled sections of the replication program the cells change into a special kind of resting state (S_0 phase), basically differing from resting in the G_0/G_1 or G_2 phases of the cell cycle. In this state they remain prepared to resume active replication immediately after stimulation by oxygenation and to retrieve the start of the sections of the replication program suspended before. Details of the real organization of sequential genome replication are not clarified so far. Figure 7 shows a conceivable hypothetical scheme (details of the depicted organization are completely arbitrary): At the beginning of the S phase a signal is generated that activates replicon clusters belonging to the first sections of the replication program. The completion of these sections generates further signals activating subsequent sections. This goes on until the total S-phase program is finished. We propose that, under hypoxia, the sections are refractory to the activating signals. Thus, cells made hypoxic in the G_1 phase (Fig. 7 A) will proceed to the point where the first replicon clusters should normally be activated and will accumulate there. Tentatively, we designate this point "pre-S-state." In S cells undergoing hypoxia (Fig. 7 B), replication will proceed to the next "signal points" and will halt there until the refractoriness is abolished after oxygen recovery. Then, the normally staggered activation of forthcoming sections will occur in a burst-like manner. However, there are indications that halting under hypoxia is not absolute and that a very slow progress through the S phase remains possible [10].

The proposed model can explain the observations, including a distraction of the S phase by repetitive deoxygenation/reoxygenation described in the next paragraph.

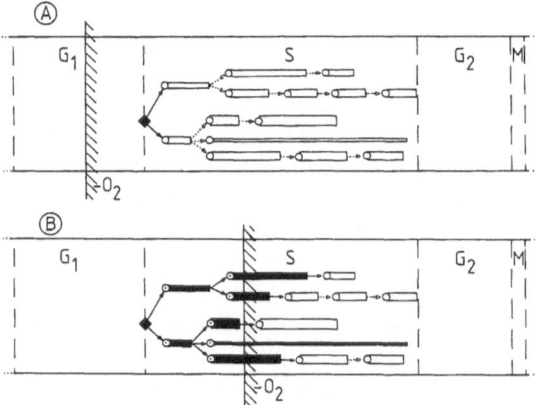

Fig. 7 A, B. Hypothetical, simplified scheme of sequential genome replication. The *bars* represent sections of a conceivable replication program. They are normally started after signals represented by *arrows*. The reception of the signals *(circles)* is refractory under hypoxic conditions. **A** State in a cell rendered hypoxic during G_1; **B** occurence of hypoxia within S

Studies on In Vivo Ascites Tumors

Relatively long ago, Harris et al. [11] suggested that the microenvironment of the cells of in vivo ascites tumors is generally hypoxic. The cells were thought to become only transiently oxygenated when fluctuating into the vicinity of blood vessels of the peritoneum. Obviously, the frequency of such transient oxygenations will diminish as tumor volume increases. On the basis of the results with cultured cells, we expected that a situation of this kind would intermittently suppress and retrigger replicon initiations, thereby causing temporal dissociation of S-phase replication. Consequently, the S-phase duration should be prolonged in vivo in comparison to permanently aerated cell culture. This was exactly what we observed earlier with our cells [7], in spite of enhanced DNA chain growth under in vivo conditions [12]. Furthermore, the S-phase duration should increase in the course of in vivo tumor growth. This is also a published fact [13]. Continuing this reasoning, we expected to see corresponding differences in the length distributions of replicative daughter strand DNA reflecting the mean relative replicon initiation frequency.

Figure 8 A demonstrates that the relative frequency of replicon initiations was indeed distinctly decreasing in the later stages of in vivo ascites tumor growth. Figure 8 B shows the drastic difference between growth as in vivo tumor and as aerated cell culture. The observed changes correspond well with the labeled mitoses curves recorded earlier under analogous conditions [7]. Figure 8 C and D shows the results of an attempt to manipulate the replicon initiation frequency of the cells during growth in the peritoneal cavity of the mice by allowing the animals to breathe elevated (Fig. 8 C) or diminished (Fig. 8 D) O_2 concentrations. As expected on the basis of the oxygen dissociation curve of mouse blood [14], the effect of increased O_2 was only marginal (but reproducible) whereas breathing of decreased O_2 concentrations caused distinct diminution of short labeled chains originating from recently initiated replicons. Thus, the results were compatible with the predictions deduced by applying the model of intermittent oxygenation, formulated by Harris et al. [11], to oxygen-dependent suppression and triggering of replicon initiations. It is expected that relatively short shutdown periods do not interrupt S-phase replication but cause execution in a more staggered manner, thus prolonging its duration. Thereby the fraction of S cells should increase rather than decrease. This was, indeed, demonstrated by the thymidine pulse labeling index. The latter was higher in 4-day-old in vivo tumors than in cell cultures prepared from the same, although cell growth was markedly enhanced in the continually aerated cultures [7]. During in vivo tumor growth the thymidine pulse labeling index remained essentially constant between the 2nd and the 7th day following inoculation (50%–52% labeled cells) and then decreased to 20%–30% between the 9th and the 12th day when the in vivo growth curve (Fig. 9, left) already distinctly flattened. However, DNA histograms obtained by flow cytofluorometry indicated no corresponding decrease of S cells (Fig. 9, right).

Probably, the decrease of the thymidine labeling index in these late stages of tumor growth reflects a marked increase of the mean period of time between two oxygenations triggering initiations of new replicon groups. Thereby, the staggered execution of S phase replication is expected to be progressively spaced by the em-

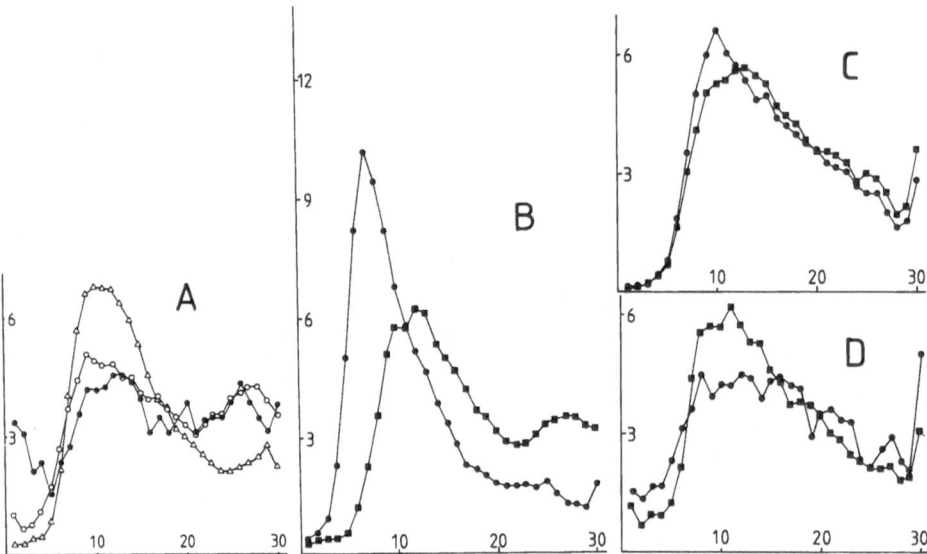

Fig. 8 A-D. Alkaline sedimentation analysis of the length distributions of growing daughter strand DNA of in vivo tumors. The DNA was labeled in vivo by i.p. injection of [³H]dThd according to [5]. **A** Tumors of different age: △, 2 days; ○, 6 days; ●, 9 days; **B-D** Effects of changing the O_2 supply of in vivo tumor cells. **B** Two mice were inoculated with the same donor ascites. After 4 days, the tumor cells of one were labeled in vivo by injection of [³H]dThd (■), that of the other in aerated culture 2 h after explantation (●). **C** One of two identically inoculated tumor mice was kept between 50 and 70 h after the inoculation in an atmosphere consisting of 96% O_2 and 4% CO_2 and then, at the same time as the control animal breathing normal air, injected i.p. with [³H]dThd; 10 min later, the tumor cells were withdrawn and analysed by alkaline sedimentation. ■, control (air); ●, 96% O_2. **D** Same schedule as **C** except that the mouse breathed 6% O_2/94% N_2 (●), while the control mouse (■) breathed normal air (from [5])

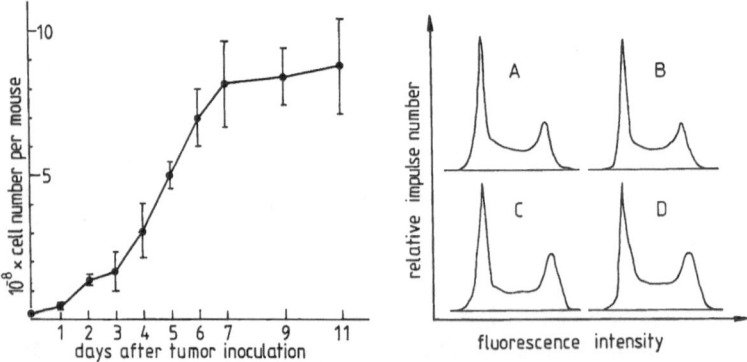

Fig. 9 A-D. *Left:* Growth curve of the ascites tumor in vivo. The *vertical bars* indicate S. D. ($n = 5$). *Right:* DNA histograms obtained by flow cytofluorometry of cells from ascites tumors of different age. **A,** 2 days; **B,** 5 days; **C,** 8 days; **D,** 12 days (from [5])

ergence of pauses during which the activity of DNA synthesis in individual cells drops below the threshold of detection by autoradiography. The volume of an ascites tumor at the end of the growth curve shown in Fig. 9 was 6–10 ml amounting to ¼–⅓ of the total body mass of the mouse. It is not surprising that oxygenation of this large volume by the peritoneal capillaries is insufficient. Finally, it should be mentioned that cells from the plateau phase of the growth curve are normally viable and grow essentially normally when explanted to culture. The effectivity as an inoculum for new ascites tumors is also not noticeably altered. We conclude from this that the plateau phase is not a result of significant cell damage which possible could occur in large tumors, but is brought about by the oxygen-dependent regulation of replication.

Application of Hypoxia as a Tool for Studying DNA Replication in Animal Cells

The rapid decrease of the initiation frequency after deoxygenation can be exploited for studying discontinuous DNA synthesis at replication forks without interference by short DNA chains occurring in very recently started replicons [15].

When selected G_1 cells were accumulated by hypoxia in the "pre-S-state" (see above and Fig. 6 C and D) the degree of synchrony of the cell population was improved (Fig. 10). It was possible to trigger, by reoxygenation, a sharp onset of S-phase replication which was followed by normal further cycling at relatively good synchrony (Fig. 10 B). Application of oxygen deprivation for preparing synchronous animal cells was already proposed earlier by Löffler et al. [16]. There are various drugs affecting replicon initiation in animal cells [17, 18, 19, 20, 21, 22, 23]. In most cases, the specificity with respect to the initiation step of replication is poor when compared to hypoxia. In our hands, only low doses of cycloheximide suppressed, like hypoxia, forthcoming initiations without noticeably affecting chain elongation and maturation [24]. Whereas 2–3 h are necessary to obtain sufficient

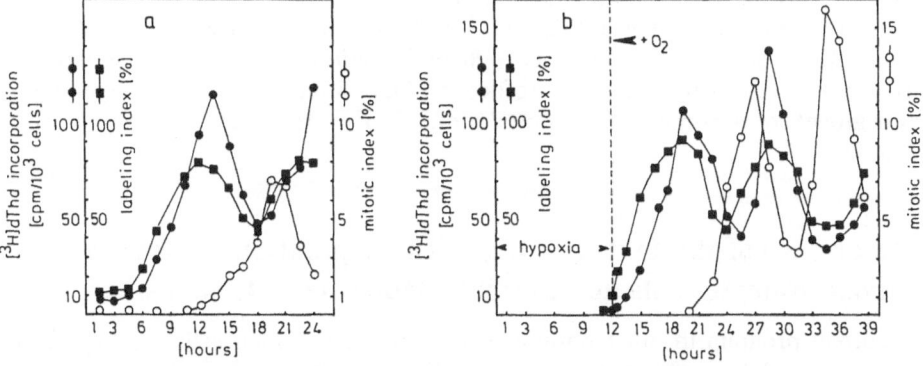

Fig. 10 a, b. DNA synthesis rate, thymidine pulse labelling index, and mitotic index during in vitro growth of G_1 cells selected by zonal centrifugation (a). A part of the same selected G_1 cells (b) was kept under hypoxic conditions for 12 h, reoxygenated, and evaluated as culture (a) (from [24])

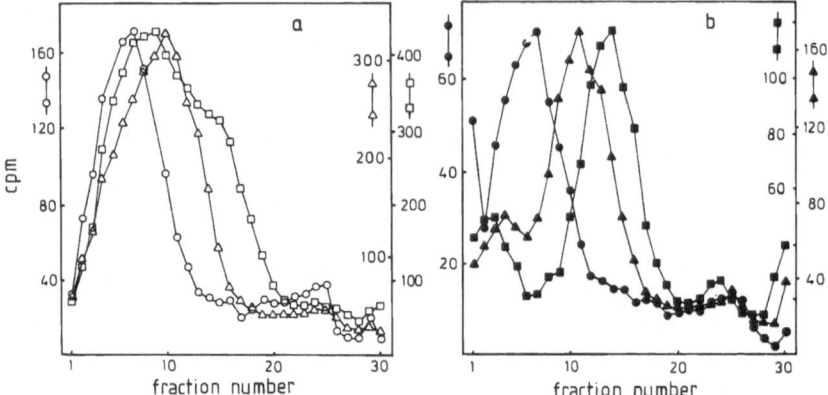

Fig. 11 a, b. Analysis of replicon growth in reaerated "pre-S-cells" by alkaline sucrose gradient centrifugation. Selected G_1 cells were reoxygenated after 12 h hypoxia. The culture was then divided; 30 μM cycloheximide were added to one part within the first min after reoxygenation. Cell samples from both cultures in part were pulse labeled for 5 min with [³H]dThd at the times indicated below, lysed on the top of alkaline sucrose gradients, and sedimented. Sedimentation direction in the depicted profiles is from *left* to *right*. One fraction corresponds to 5.4 S. **a** Control without cycloheximide, labeling 5 min after reoxygenation —O—O—; 1 h after reoxygenation —△—△—; 2 h after reoxygenation —□—□—. **b** Culture with cycloheximide, time of labeling: —●—●— 5 min; —▲—▲— 1 h; —■—■— 2 h after reoxygenation (from [24]

O_2 deprivation of cell cultures, cycloheximide has proved to effect strong initiation suppression within a few minutes. However, cycloheximide was distinctly inferior to hypoxia with respect to reversibility after drug removal. Exploiting the excellent reversibility of the hypoxia mediated shutdown of initiation and the quick action of cycloheximide, we elaborated a protocol for preparing cell populations, in which exclusively the very first replicons of the S phase acted synchronously. This protocol consists of: (a) Selection of G_1 cells; (b) accumulation in the "pre-S-state" by 12 h hypoxia; (c) reoxygenation; and (d) addition of 30 μM cycloheximide within the first minute after reoxygenation. Figure 11 shows the alkaline sedimentation profiles of growing daughter strand DNA obtained in this type of experiment. The differences between Figs. 11 A and B clearly demonstrate that cycloheximide effectively suppressed further replicon initiations after a first burst had been triggered by reoxygenation. The profiles in Fig. 11 B reflect synchronous initiation and growth of a homogeneous population of DNA molecules to replicon sized lengths.

Occurrence of the Oxygen-Dependent Regulation of Replication Among Animal Cells and Possible Molecular Mechanisms

A current program in our laboratory will evaluate in which cells the oxygen-dependent regulation of replication occurs. Previous data indicate that it is not restricted to Ehrlich ascites cells. In some cell lines (e.g., BHK cells) we observed effects even more distinct than in the Ehrlich ascites cells. Other cell lines, especially tumor cells growing very slowly, did not respond in the expected manner. More data have to be collected to clarify this point. Our search for an oxygen sensor has been unsuccessful so far. The action of cycloheximide points towards a possible involvement of a short-lived protein in the finally observed on/off switching of replicon initiation. There are further data [25, 26] suggesting a possible role of the deoxycytidine triphosphate level in the intracellular transmission of the oxygen signal. More research will be necessary to elucidate the molecular mechanism of the oxygen-dependent regulation of replication in animal cells.

Acknowledgements. The authors are grateful to Dr. Gudrun Probst for aid in preparing the manuscript. The work of the authors was supported by the Deutsche Forschungsgemeinschaft (Pr 95/10-1)

References

1 Hand R (1978) Eukaryotic DNA: organisation of the genome replication. Cell 15: 317–325
2 Taylor JH (1984) Origins of replication and gene regulation. Mol Cell Biochem 61: 99–109
3 Probst H, Gekeler V (1980) Reversible inhibition of replicon initiation in Ehrlich ascites cells by anaerobiosis. Biochem Biophys Res Commun 94: 55–60
4 Probst H, Gekeler V, Helftenbein E (1984) Oxygen dependence of nuclear DNA replication in Ehrlich ascites cells. Exp Cell Res 154: 327–341
5 Probst H, Schiffer H, Gekeler V, Kienzle-Pfeilsticker H, Stropp U, Stötzer KE, Frenzel-Stötzer I (1988) Oxygen dependent regulation of DNA synthesis and growth of Ehrlich ascites tumor cells. Cancer Res (in press)
6 Huberman JA, Riggs AD (1968) DNA replication in mammalian chromosomes. J Mol Biol 32: 327–341
7 Probst, H, Maisenbacher, J (1973) Use of zonal centrifugation for preparing synchronous cultures from Ehrlich ascites cells grown in vivo. Exp Cell Res 78: 335–344
8 Froese G (1962) The respiration of ascites tumor cells at low oxygen concentrations. Biochim Biophys Acta 57: 509–519
9 Löffler M (1984) On the role of dihydroorotate dehydrogenase in growth cessation of Ehrlich ascites tumor cells cultured under oxygen deficiency. Eur J Biochem 107: 207–215
10 Merz R, Schneider F (1982) Growth characteristics of in vitro cultured Ehrlich ascites tumor cells under anaerobic conditions and after reaeration. Z Naturforsch [37 C] 326–334
11 Harris JW, Meyskens F, Patt HM (1970) Biochemical studies of cytokinetik changes during tumor growth. Cancer Res 30: 1937–1946
12 Probst H, Blütters R, Fielitz J (1980) DNA replication in asynchronous and synchronous Ehrlich ascites cells in different conditions of growth. Exp Cell Res 130: 1–13
13 Lala PK (1971) Studies on tumor cell population kinetics. In: Busch H (ed), Methods in cancer research, vol 6. Academic, New York, pp 3–95
14 Schmidt-Neilsen K, Larimer JL (1958) Oxygen dissociation curves of mammalian blood in relation to body size. Am J Physiol 212: 424–428
15 Gekeler V, Stropp U, Probst H (1986) Application of hypoxia-induced shut down of replicon initiation to the analysis of replication intermediates in Ehrlich ascites cells. Biol Chem Hoppe-Seyler 367: 1209–1217
16 Löffler M, Postius S, Schneider F (1978) Anaerobiosis and oxygen recovery: changes in cell cycle distribution of Ehrlich ascites tumor cells grown in vitro. Virchows Arch [B] 26: 359–368
17 Fridland A (1977) Inhibition of deoxyribonucleic acid chain initiation: a new mode of action for 1-β-D-arabinofuranosylcytosine in human lymphoblasts. Biochemistry 16: 5308–5312
18 'Gautschi JR, Kern RM, Painter RB (1973) Modification of replicon operation in HeLa cells by 2,4-dinitrophenol. J Mol Biol 80: 393–403
19 Guy AL, Taylor JH (1978) Actinomycin D inhibits initiation of DNA replication in mammalian cells. Proc Natl Acad Sci USA 75: 6088–6092

20 Mattern MR, Painter RB (1979) Dependence of mammalian DNA replication on DNA super-coiling. II. Effects of novobiocin on DNA synthesis in chinese hamster ovary cells. Biochim Biophys Acta 563: 306–312
21 Kaufmann WK, Schwartz JL (1981) Inhibition of replicon initiation by 12-o-tetradecanoyl-phorbol-13-acetate. Biochem Biophys Res Commun 103: 82–89
22 Painter RB (1978) Inhibition of DNA replicon initiation by 4-nitroquinoline 1-oxide, adriamy-cin, and ethyleneimine. Cancer Res 38: 4445–4449
23 Hand R (1975) DNA replication in mammalian cells: altered pattern of initiation during inhi-bition of protein synthesis. J Cell Biol 67: 761–773
24 Gekeler V, Probst H (1988) Synchronization of replicons in Ehrlich ascites cells. Exp Cell Res 175: 97–108
25 Löffler M, Schimpff-Weiland G, Follmann H (1983) Deoxycytidylate shortage is a cause of G_1 arrest of ascites tumor cells under oxygen deficiency. FEBS Lett 156: 72–76
26 Probst H, Kienzle-Pfeilsticker H, Schiffer H, Gekeler V (in preparation) Triggering of replicon initiations in hypoxic Ehrlich ascites cells by deoxycytidine (in preparation).

Metabolic Events that May Activate Erythropoietin Production in the Hypoxic Kidney

C. Bauer

Physiologisches Institut der Universität Zürich, Winterthurerstraße 190, 8057 Zürich, Switzerland

Introduction

Hypoxia is a condition that is poorly defined in any sense of the word. Nevertheless, there is an agreement among physiologists that hypoxia represents a state of reduced oxygen availability that causes measurable changes in certain physiological parameters. That is to say that a hypoxic state is inferred from the adaptive reaction rather than from a definite critical oxygen tension below which a given tissue is considered hypoxic. This ex reactio definition implies that there is a broad range of oxygen tensions through which an organ can adjust its biological activities to compensate for suboptimal oxygen supply.

Notable examples for such compensatory mechanisms that occur at suboptimal oxygen supply are stimulation of respiration via peripheral chemoreceptors, increase in cardiac output, rise in coronary and cerebral blood flow, and stimulation of erythropoietin production in the kidney. In the following brief overview I will consider the events that may lead to an enhanced production of erythropoietin in the kidney when there is a disproportion between renal oxygen supply and renal oxygen consumption. In its entirety, the metabolic events that are set in motion under such hypoxic conditions may be called a renal oxygen sensor. I will use a broad definition of oxygen sensor mechanisms by which changes in oxygen tension result in an effector mechanism, e.g., increased formation of erythropoietin. After giving some background on erythropoietin, I will consider the changes in cellular metabolism that may occur under hypoxic conditions and then turn to those events that may be directly involved in the activation of the machinery that manufactures erythropoietin.

Background on Erythropoietin Structure and Function

Erythropoietin is a glycoprotein hormone with a relative molecular mass of 34000 daltons of which the carbohydrate part represents some 40%. One can speculate that the hydrophilic, oligosaccharide structures are required for maintenance of the conformation of the hydrophobic protein, but may not be directly involved in the interaction with cellular receptors [20, 21, 22]. The amino acid sequence of erythropoietin from man, monkey, and mouse have a very high degree of homology (85% on the average) and are thus very highly conserved primary structures [26, 37, 39, 40, 44, 57]. The main function of erythropoietin is the stimulation of red blood cell formation in the bone marrow. It does so by enhancing mitosis of certain pools of erythroid precursor cells. These pools are mainly functionally defined, i.e., by their responsiveness towards erythropoietin and by the synthesis of a series of characteristic red blood cell proteins [25, 48, 49].

The normal concentration of erythropoietin in the blood plasma is in the picomolar range. An increase of the concentration of erythropoietin will result in an increase in red cell count that helps to alleviate the effects of hypoxia on the body. The gene that codes for erythropoietin has been cloned for mice, monkeys, and humans [26, 38, 39, 40, 44, 57]. By using Northern blot techniques for the detection of erythropoietin mRNA it became possible to unequivocally localize the kidney as the main production site of the hormone [4, 5]. The use of molecular probes confirmed therefore what has long been conjectured from physiological experimentation where removal of the kidney almost completely suppressed hypoxia-induced production of erythropoietin [27]. In addition, recombinat human erythropoietin became available that can be used for the treatment of patients who suffer from severe anemia due to end-stage renal disease [14, 67].

The Oxygen Sensor in the Kidney

The major and physiologically most important stimulus for an increased production of erythropoietin in the kidney is a hypoxic condition. As was mentioned in the Introduction, hypoxia is poorly defined and I will use it here to indicate a deficiency of oxygen that is severe enough to lead to the elaboration of the hormone. Such a deficiency of oxygen can be brought about in a number of conditions, e.g., decrease in arterial oxygen tension, decresae in red cell count, carbon monoxide poisoning, and stenosis of the renal artery [16, 24, 28, 30, 41].

In view of the fact that such a diversity of different hypoxic states all lead to an increased production of erythropoietin, one might wonder what kind of cellular signals are being generated when oxygen becomes scarce. In the following paragraph, evidence will be furnished to show that ATP-dependent reactions can, in principle, serve as oxygen sensors or oxygen chemoreceptors.

The ATP-Dependent Reactions

A reduced availability of oxygen will, at least at the cellular level, lead to a decrease in oxygen consumption only at very low tensions of oxygen. Of course, in the intact organ the decrease in oxygen consumption occurs at higher oxygen tensions than in isolated cell preparations. Thus, in the isolated perfused rat kidney with no red cells added to the perfusate, the venous pO_2 below which oxygen consumption decreases is around 30–40 torr [13]. In order to achieve such an independence of oxygen consumption on oxygen availability over a relatively large range of oxygen tensions, certain metabolic adaptations have to occur. Wilson and coworkers have shown in preparations of isolated cells that the extent of cytochrome c reduction depends on the oxygen tension at all values measured up to and including that of media saturated with air [66]. At constant oxygen consumption of the cells, the redox state of cytochrome c is linked to changes in the mitochondrial [NAD$^+$]/[NADH] and the cytosolic [ATP]/[ADP][Pi]. From the equations Wilson and colleagues derived it is evident that any reduction of cytochrome c must be accompanied either by a reduction of the mitochondrial NAD$^+$ pool or by a decrease in cytosolic [ATP]/[ADP][Pi] or both [65, 66]. It was observed experimentally that the cellular [ATP]/[ADP][Pi] behaved almost as mirror image of reduced cytochrome c at oxygen tensions greater than approximately 3 torr, which suggests that there is little change in mitochondrial [NAD$^+$]/[NADH] until very low oxygen tensions are reached. The dependence of cytosolic [ATP]/[ADP][Pi] on oxygen availability meets the requirements for a metabolic "sensor" of cellular oxygen tension [66].

I will now discuss some possible mechanisms by which this signal is amplified and translated to a physiological effector mechanism. I will restrict myself on the role of ATP even though ADP or Pi are certainly candidates for exerting an allosteric control on certain effector molecules, like the activation of phosphofructokinase by ADP or the inhibition of the Na$^+$ pump by ADP and Pi [56, 58]. However, in the present context my emphasis is on the consideration of possible ATP-dependent reactions and how they might be coupled to physiological effector mechanisms. It should be pointed out here, that ATP-dependent reactions involved in the signalling of oxygen should have low ATP affinities whilst ATP-dependent reactions essential to cell survival should have a high affinity for ATP. Thus in the former case (oxygen chemoreception), a small decrease in the concentration of ATP could initiate a specific cellular reaction, while in the latter case (essential cell reactions) the ATP concentrations must fall to very low levels before cell survival is being endangered [8, 54, 62]. Therefore, the specificity of an oxygen chemoreceptor would reside in the specific set of functions that are regulated by low-affinity ATP binding sites.

It is worth remembering in this connection that compartmentation is still another and very possibly complementary way to achieve specificity in cellular functions. In smooth muscle cells, for example, glycolytic and oxidative metabolism are functionally compartmentalized with a close coupling of glycolysis and Na$^+$-K$^+$-ATPase [43, 51, 52, 53]. The same phenomenon is observed in heart cells [19, 45] and human red cells [46]. Another specialization of ATP-dependent reactions was described by Aw and Jones [1]. These investigators have studied the activities of two ATP-requiring systems with different subcellular localizations in rat hepatocytes in which average cellular ATP concentration was varied. The cytosolic ATP-sulfurylase activity varies linearly with the cellular ATP concentration, however, the plasma membrane Na$^+$-K$^+$-ATPase is substantially more sensitive to decreased ATP concentration compared to the ATP-sulfurylase activity. These results indicate that ATP-dependent enzymes located in the plasma membrane in liver cells are exposed to a lower ATP concentration than are enzymes in the cytosolic fluid around the mitochondria. Calculations using previously determined ATP diffusion coefficients and an assumed geometry, suggest that radial ATP gradients may occur during hypoxia so that ATP levels at the cytosolic face of the plasma membrane will reach nearly zero [32]. During normoxia, however, ATP gradients are minimal. These data suggest a basis for considering plasma membrane ATP-dependent reactions linked to oxygen chemoreception. There are three examples that may be relevant in this respect: (a) the ATP-dependent K^+-channel; (b) the phosphatidylinositol phosphate kinase (PIP kinase); and (c) the acyl-CoA synthetase. Examples (a) and (b) will be dealt with only briefly because they have no known relationship with the mechanisms that lead to an enhanced erythropoietin formation in the kidney. Example (c) will be treated in some more detail because of its possible relevance to the oxygen sensor in the kidney.

ATP-Dependent K^+ Channels

Patch-clamp techniques reveal ATP-dependent K^+ channels in cardiac muscle [33, 50] and in pancreatic B cells [11]. Treatment with cyanide or perfusion of the cytoplasmatic surface of the membrane with ATP-free solutions results in increases in open times and in frequency of open times of these delayed rectifying channels which are insensitive to membrane potential and intracellular pH [11]. The apparent K_m for ATP for this effect is about 0.2 mM at the cytoplasmic face of the plasma membrane from myocardial cells under patch-clamp conditions [50]. Remember, however, that diffusion gradients for ATP may exist in an intact cell under hypoxic conditions [32] so that the effective K_m for ATP under physiological conditions may be substantially higher. This is reminiscent of the ATP dependence of the renal Na^+ pump which has a linear, nonsaturating dependence on ATP concentration in isolated renal proximal tubules. However, the Na^+-K^+-ATPase hydrolytic activity of lysed tubular membranes demonstrates saturation and has a K_m value of 0.4 mM ATP [58].

The characteristics of ATP-dependent K^+ currents in cardiac muscle and pancreatic B cells referred to above seem to be qualitatively similar to those of the voltage-independent K^+ currents which result in the hypoxia-induced hyperpolarization in rat hippocampal neurones in vitro [18].

PIP Kinase

Lundberg and colleagues [42] have shown that phosphorylation of phosphatidylinositol phosphate (PIP) to phosphatidylinositol bisphosphate (PIP$_2$) was inhibited in hepatocyte membrane preparations when the ATP concentration fell below 1 mM. The reaction by which PIP is phosphorylated to PIP$_2$ is part of a series of reactions which transduce and amplify signals generated by receptor binding to agonist, and provide second messengers like inositol triphosphate and diacylglycerol [64]. Thus, the attractiveness of PIP kinase functioning as an ATP-sensor during oxygen chemoreception reactions includes the high K_m for ATP and a central role of the inositol phospholipid transduction system in control of secretion, contractility, and other cellular responses [9]. The hypothesis that PIP kinase can function as an ATP-sensor during oxygen chemoreception was experimentally tested by Coburn and his coworkers [10]. These investigators measured contractility and phosphatidylinositol metabolism in the isolated rabbit aorta incubated in an organ bath at normoxia and hypoxia (pO$_2$ < 40 torr). Coburn et al. have found that tissue pool size of PIP when expressed as total pool of PIP/total pool of PIP$_2$

increases by 300% when the tissue is activated by norepinephrine during hypoxia compared to normoxia. Under hypoxia there appears to be a decreased ATP synthesis and delivery to plasma membrane PIP kinase, and possibly to ATP-dependent reactions involved in PI resynthesis causing a decrease in metabolic flux in inositol phospholipid and a decrease in the synthesis of inositol phosphates like IP, IP_2, and IP_3 [10].

Acyl-CoA Synthetase

Morita and Murota [47] have shown in a preparation of rat liver microsomes that ATP shifts arachidonic acid away from cyclooxygenase into phospholipids. Their finding is important because it suggests that the availability of arachidonic acid for prostaglandin synthesis via the cyclooxygenase reaction is ATP-dependent. Arachidonic acid, as will be remembered, is mostly esterified to the hydroxyl group at carbon atom 2 of phosphoglycerides. Therefore, an increse in the availability of arachidonic acid can be achieved either by an activation of hydrolysases (phospholipases) that cleaves arachidonic acid from its ester bonds, or by an inhibition of reesterification of arachidonic acid into the phospholipid pools. An increased supply of arachidonic acid to the cyclooxygenase will obviously lead to an increased synthesis of prostaglandin. At this point erythropoietin regulation comes into play. In fact, there are numerous experiments described in the literature which show that a functioning prostaglandin system is necessary for the elaboration of hypoxia-induced erythropoietin in vivo, in the isolated perfused kidney and in cell cultures [12, 15, 24, 34, 55]. It has furthermore been shown, that arachidonic acid and prostaglandins of the E series can stimulate erythropoietin production in the isolated perfused kidney [17, 23, 28] and in cell cultures that produce erythropoietin [34]. Based on these results, the hypothesis has been put forward that prostaglandin E_2 is one signal molecule that is generated in the kidney upon hypoxia and can stimulate erythropoietin production [15]. If this is really the case, it is necessary to demonstrate an increase of prostaglandin release under hypoxic conditions. Indeed, hypoxia increases the release of prostaglandins from the kidney under in vivo conditions [60, 63] as well as from cultured renal cells [31].

The question that arises from these findings is related to the mechanism by which hypoxia leads to an increased prostaglandin release. We have addressed this problem by investigating arachidonic acid and prostaglandin metabolism in a permanent renal tubular cell line, the MDCK cells, that can be stimulated to transport more NaCl from the basolateral side to the apical side. Under these stimulated conditions oxygen consumption increases by a factor of two [35] and the peak release of prostaglandins and lactate increases by factors of 30 and 10, respectively [36]. When the cells are prelabelled with ^{14}C arachidonic acid and again stimulated to increase NaCl transport and oxygen consumption, we found that the lipids from which the radioactivity is liberated correspond mostly to PIP and PIP_2 [36].

The next question regards the mechanisms by which arachidonic acid is released from the phospholipids. As was stated above, there are two possibilities by which the delivery of arachidonic acid to the cyclooxygenase can be enhanced: Increased hydrolysis due to the activation of phospholipases or decreased reesterification due to the inhibition of acyl-CoA synthetase. Our results [35, 36] favor the view that under the conditions employed it is not the increased hydrolysis but rather the inhibition of an acyl-CoA synthetase that leads to an enhancement of prostaglandin synthesis. Three lines of evidence argue in favor of this possibility: (a) Long chain fatty acyl-CoA synthetase hs a very high K_m value for ATP, namely about 4.5 mM [2]. (b) Graded depletion of ATP from the MDCK cells by using rotenone and amobarbital results in an enhanced production of prostaglandin E_2 [36]. (c) Hypoxia leads to the steepening of an ATP gradient between the mito-chondria and the periphery of the cell [1]. Therefore, under such hypoxic condi-tions the cellular ATP pools become compartmentalized with a fall of the ATP concentration at those cellular localizations that have a high ATP consumption and are far away from the mitochondrial production sites or both.

In summary, the release of prostaglandins under hypoxic conditions appears to be due to an inhibition of the reesterification of arachidonic acid, which in turn is caused by a reduced supply of ATP to a long chain fatty acyl-CoA synthetase that has an estimated K_m for ATP of about 4.5 mM, i.e., a value that is near the normal intracellular concentration of ATP.

The prostaglandins that are being released by this mechanism can cause a number of reactions in the kidney: they would promote vasodilatation [61, 68] and would also down-regulate transport activity of the medullary thick ascending limb [59]. In addition, there is ample evidence that prostaglandins, particularly of the E-type, stimulate erythropoietin production both in vivo [15, 55] and in cell cultures [34]. The increase in erythropoietin formation is very likely caused by the activa-tion of the adenylate cyclase that leads to the activation of the gene coding for erythropoietin. It should be noted that the normoxic kidney does not contain sig-nificant amounts of stored erythropoietin [29] and that the hypoxia-induced in-crease in erythropoietin production appears to be regulated exclusively by increas-ing the rate of transcription [5, 44]. Therefore, the cells that are involved in the elaboration of erythropoietin belong to the ones that are being stimulated by cyclic AMP to produce a specific product [3], in this case a hormone that counteracts hypoxia by stimulating red blood cell formation.

Conclusion

I have presented evidence to show that a fall of the ATP concentration at strategi-cally important points within a cell is one possible way by which the transduction mechanism of a renal oxygen sensor could activate certain effector mechanisms, e.g., increased prostaglandin synthesis. Within the framework of such a sequence of metabolic reactions an important link would be an ATP-dependent enzyme that has a high K_m-value for ATP. A slight fall in the concentration of ATP caused by hypoxia could therefore initiate protective mechanisms that are geared to alleviate the effects of hypoxia, e.g., increased blood flow, decreased ATP consumption [6, 7], or an increase of the oxygen-carrying capacity of the blood. The first two mech-anisms would have a short time requirement and act mainly locally. An increase in erythropoietin production with a consecutive rise in the oxygen-carrying capacity of the blood takes about 1 week to become effective, and is an adaptive mechanism in conditions of chronic hypoxia. In either case, one has to postulate that the mechanisms that are activated under hypoxic conditions never fully compensate a given degree of hypoxia. Only when there is a constant "backlog" in the regulation can the compensatory mechanisms remain activated as long as the cause for hypoxia remains present.

Acknowledgements. I am grateful to Ronald Coburn, Ulrich Pohl, and Armin Kurtz for very helpful and stimulating discussions. The research done in the author's laboratory is supported by the Swiss National Science Foundation, the Roche Research Foundation, and the Hartmann Müller Stiftung für medizinische Forschung

References

1 Aw TY, Jones DP (1985) ATP concentration gradients in cytosol of liver cells during hypoxia. Am J Physiol 249: C385-C392
2 Bar-Tana J, Rose G, Shapiro B (1975) Long chain fatty acyl-CoA synthetase from rat liver microsomes. In: Lowenstein JM (ed) Methods of enzymology vol 35. Academic, New York, pp 117-122
3 Bauer C (1987) Chemoreception of oxygen in the kidney and erythropoietin production. In: Rich IN (ed) Molecular and cellular aspects of erythropoietin and erythropoiesis. NATO ASI series vol H 8. Springer, Berlin Heidelberg New York, pp 311-327
4 Beru N, McDonald J, Lacombe C, Goldwasser E (1986) Expression of the erythropoietin gene. Mol Cell Biol 6: 2571-2575
5 Bondurant MC, Koury MJ (1986) Anemia induces accumulation of erythropoietin mRNA in the kidney and liver. Mol Cell Biol 6: 2731-2733
6 Brezis M, Rosen S, Silva P, Epstein FH (1984) Renal ischemia: a new perspective. Kidney Int 26: 375-383
7 Brezis M, Rosen S, Silva P, Epstein FH (1984) Selective vulnerability of the thick ascending limb to anoxia in the isolated perfused rat kidney. J Clin Invest 73: 182-190
8 Carafoli E (1984) Calmodulin-sensitive calcium-pumping ATPase in plasma membranes: isolation, reconstitution, and regulation. Fed Proc 43: 3005-3010
9 Coburn RF (1988) ATP-sensing reactions and oxygen chemoreception. In: S Lahiri (ed) Receptors and reflexes in breathing. Oxford University Press, New York (in press)
10 Coburn RF, Baron C, Papadopoulos MT (1987) Phosphoinositide metabolism in rabbit aorta is altered during hypoxia. (submitted for publication)
11 Cook DL, Hales N (1984) Intracellular ATP directly blocks K^+ channels in pancreatic B-cells. Nature 311: 271-273
12 Dukes PP, Shore NA, Hammond D, Ortega JA, Datta MC (1973) Enhancement of erythropoiesis by prostaglandins. J Lab Clin Med 82: 704-712
13 Dume T, Koch KM, Krause HH, Ochwadt B (1966) Kritischer venöser Sauerstoffdruck an der erythrocytenfrei perfundierten isolierten Rattenniere. Pflugers Arch 290: 89-100
14 Eschbach JW, Egrie JC, Downing MR, Browne JK, Adamson JW (1987) Correction of the anemia of end-stage renal disease with recombinant human erythropoietin. N Engl J Med 316: 73-78
15 Fisher JW (1980) Prostaglandins and kidney erythropoietin production. Nephron 25: 53-56
16 Fisher W, Schofield R, Porteous DD (1965) Effects of renal hypoxia on erythropoietin production. Br J Haematol II: 382-388
17 Foley JW, Gross DM, Nelson PK, Fisher JW (1978) The effects of arachidonic acid on erythropoietin production in exhypoxic polycythemic mice and the isolated perfused canine kidney. J Pharmacol Exp Ther 207: 402-409
18 Fujiwara N, Higashi H, Shimoji K, Yoshimura M (1987) Effects of hypoxia on rat hippocampal neurones in vitro. J Physiol (Lond) 384: 131-151

19 Girardier L (1971) Dynamic energy partition in cultured heart cells. Cardiology 56: 88-92
20 Goldwasser E, Kung CK-H, Eliason JF (1975) On the mechanism of erythropoietin-induced differentiation XIII. The role of sialic acid in erythropoietin action. J Biol Chem 249: 4202-4206
21 Goldwasser E, Eliason JF, Sikkema D (1975) An assay for erythropoietin in vitro at the milliunit level. Endocrinology 97: 315-323
22 Goldwasser E, McDonald J, Beru N (1987) The molecular biology of erythropoietin and the expression of its gene. In: Rich IN (ed) NATO ASI Series H8. Molecular and cellular aspects of erythropoietin and erythropoiesis. Springer, Berlin Heidelberg New York, pp 11-21
23 Gross DM, Fisher JW (1980) Erythropoietic effects of PGE_2 and 2 endoperoxide analogs. Experientia 36: 458-459
24 Gross DM, Mujovic VM, Jubiz W, Fisher JW (1976) Enhanced erythropoietin and prostaglandin E production in the dog following renal artery constriction. Proc Soc Exp Biol Med 151: 498-501
25 Harrison PR, Frampton J, Chambers I, Kasturi K, Thiele B, Conkie D, Fleming J, Chester J, O'Prey J, McBain W (1987) Analysis of erythroid cell-specific gene expression. In: Rich IN (ed) NATO ASI series vol H8. Molecular and cellular aspects of erythropoietin and erythropoiesis. Springer, Berlin Heidelberg New York, pp 37-50
26 Jacobs , Shoemaker C, Rudersdorf R, Neill EF, Kaufman RJ, Mufson A, Seehra J, Jones SS, Hewick R, Fritsch EF, Kawakita M, Shimaza T, Miyake T (1985) Isolation and characterization of genomic and cDNA clones of human erythropoietin. Nature 313: 806-810
27 Jacobson LO, Goldwasser E, Fried W, Plzak LF (1957) The role of the kidney in erythropoiesis. Nature 179: 633-634
28 Jelkmann W (1986) Renal erythropoietin: properties and production. Rev Physiol Biochem Pharmacol 104: 140-215
29 Jelkmann W, Bauer C (1981) Demonstration of high levels of erythropoietin in rat kidneys following hypoxic hypoxia. Pflugers Arch 393: 34-39
30 Jelkmann W, Kurtz A, Seidl J, Bauer C (1984) Mechanisms of the renal glomerular erythropoietin production. In: Grote J, Witzleb E (eds) Atemgaswechsel und O_2-Versorgung der Organe. Akademie der Wissenschaften und der Literatur, Mainz, pp 130-137
31 Jelkmann W, Kurtz A, Förstermann U, Pfeilschifter J, Bauer C (1985) Hypoxia enhances prostaglandin synthesis in renal mesangial cell cultures. Prostaglandins 30: 109-118
32 Jones DP (1986) Intracellular diffusion gradients of O_2 and ATP. Am J Physiol 250: C663-C675
33 Kakei M, Noma A (1984) Adenosine-5-triphosphate-sensitive single potassium channel in the atrioventricular node cell of the rabbit heart. J Physiol (Lond) 352: 265-284
34 Kurtz A, Jelkmann W, Pfeilschifter J, Bauer C (1985) Role of prostaglandins in hypoxia-stimulated erythropoietin production. Am J Physiol 249: C3-C8
35. Kurtz A, Pfeilschifter J, Brown CDA, Bauer C (1986) NaCl transport stimulates prostaglandin release in cultured renal epithelial (MDCK) cells. Am J Physiol 250: C676-C681
36 Kurtz A, Pfeilschifter J, Malmström K, Woodson RD, Bauer C (1987) Mechanism of NaCl transport-stimulated prostaglandin formation in MDCK cells. Am J Physiol 252: C307-C314
37 Lai P-H, Everett R, Wang F-F, Arakawa T, Goldwasser E (1986) Structural characterization of human erythropoietin. J Biol Chem 261: 3116-3121
38 Lin F-K (1987) The molecular biology of erythropoietin. In: Rich IN (ed) NATO ASI series, vol H8. Molecular and cellular aspects of erythropoietin and erythropoiesis. Springer, Berlin Heidelberg New York
39 Lin F-K, Suggs S, Lin C-H, Browne J, Smalling R, Egrie J, Chen K, Fox G, Martin F, Stabinsky Z, Badrawi S, Lai P-H, Goldwasser E (1985) Cloning and expression of the human erythropoietin gene. Proc Natl Acad Sci USA 82: 7580-7584
40 Lin F-K, Lin CH, Lai PH, Browne JK, Egrie JC, Smalling R, Fox GM, Chen KK, Castro M, Suggs S (1986) Monkey erythropoietin gene: cloning, expression and comparison with the human erythropoietin gene. Gene 44: 201-209
41 Luke RG, Kennedy AC, Stirling WB, McDonald GA (1965) Renal artery stenosis, hypertension and polycythaemia. Br Med J [Clin Res] 1: 164-166
42 Lundberg GA, Jergil B, Sundler R (1985) Subcellular localization and enzymatic properties of rat liver phosphatidylinositol-4-phosphate kinase. Biochim Biophys Acta 846: 379-387

43 Lynch RM, Paul RJ (1983) Compartmentation of glycolytic and glycogenolytic metabolism in vascular smooth muscle. Science 222: 1344-1346
44 McDonald JD, Lin F-K, Goldwasser E (1986) Cloning, sequencing, and evolutionary analysis of the mouse erythropoietin gene. Mol Cell Biol 6: 842-848
45 McDonald TF, Hunter EG, MacLeod DP (1971) Adenosinetriphosphate partition in cardiac muscle with respect to transmembrane electrical activity. Pflugers Arch 322: 95-108
46 Mercer RW, Dunham PB (1981) Membrane-bound ATP fuels the Na/K pump. Studies on membrane-bound glycolytic enzymes on inside out vesicles from human red cell membranes. J Gen Physiol 78: 547-568
47 Morita I, Murota S-I (1980) Prostaglandin-synthesizing system in rat liver homogenates: ATP shifts arachidonic acid away from cyclooxygenase into phospholipids. Biochim Biophys Acta 619: 428-431
48 Nijhof W, Wierenga PK (1983) Isolation and characterization of the erythroid progenitor cell: CFU-E. J Cell Biol 96: 386-392
49 Nijhof W, Wierenga PK (1987) The purification of spleen CFJU-E and its applications in the study of in vitro erythropoiesis. In: Rich IN (ed) NATO ASI series vol H8. Molecular and cellular aspects of erythropoietin and erythropoiesis. Springer, Berlin Heidelberg New York, pp 73-87
50 Noma A (1985) ATP-regulated single K channels in cardiac muscle. Nature 305: 147-148
51 Paul RJ (1983) Functional compartmentalization of oxidative and glycolytic metabolism in vascular smooth muscle. Am J Physiol 244: C399-C409
52 Paul RJ, Bauer M, Pease W (1979) Functional compartmentalization of oxidative and glycolytic metabolism in vascular smooth muscle. Am J Physiol 244: C399-C409
53 Paul RJ, Bauer M, Pease W (1979) Vascular smooth muscle: aerobic glycolysis linked to sodium and potassium transport processes. Science 206: 1414-1416
54 Persechini A, Mrwa U, Hartshorne DJ (1981) Effect of phosphorylation on the actin-activated ATPase activity of myosin. Biochem Biophys Res Commun 98: 800-805
55 Schooley JC, Mahlmann LJ (1971) Stimulation of erythropoiesis in plethoric mice by prostaglandins and its inhibition by antierythropoietin. Proc Soc Exp Biol Med 138: 523-524
56 Scruttan MC, Utter MF (1968) The regulation of glycolysis and gluconeogenesis in animal tissues. Annu Rev Biochem 37: 249-302
57 Shoemaker CB, Mistock LD (1986) Murine erythropoietin gene: cloning expression and human gene homology. Mol Cell Biol 6: 849-858
58 Soltoff SP, Mandel LJ (1984) Active ion transport in the renal proximal tubule. J Gen Physiol 84: 643-662
59 Stokes JB (1979) Effect of prostaglandin E_2 on chloride transport across the rabbit thick ascending limb of Henle. J Clin Invest 64: 495-502
60 Walker BR (1982) Diuretic response to acute hypoxia in the conscious dog. Am J Physiol 243: F440-F446
61 Walker LA, Frölich JC (1987) Renal prostaglandins and leukotrienes. Rev Physiol Biochem Pharmacol 107: 2-72
62 Walsh M, Dabrowska R, Hindrens S, Hartshorne DJ (1982) Calcium ion-independent myosin light chain kinase of smooth muscle. Biochemistry 21: 1919-1925
63 Weismann DN (1981) Altered renal hemodynamic and urinary prostaglandin response to acute hypoxemia after inhibition of prostaglandin synthesis in the anesthetized dog. Circ Res 48: 632-640
64 Williamson JR (1986) Inositol lipid metabolism and intraclelular signaling mechanisms. News Physiol Sci 1: 72-76
65 Wilson DF, Erecinska M, Drown C, Silver IA (1977) Effect of oxygen tension on cellular energetics. Am J Physiol 2 (3): C135-C140
66 Wilson DF, Erecinska M, Drown C, Silver IA (1979) The oxygen dependence of cellular energy metabolism. Arch Biochem Biophys 195: 485-493
67 Winearls CG, Oliver DO, Pippard MJ, Reid C, Downing MR, Cotes PM (1986) Effect of human erythropoietin derived from recombinant DNA on the anaemia of patients maintained by chronic haemodialysis. Lancet II: 1175-1177
68 Yared A, Kon V, Ichikawa I (1985) Mechanism of preservation of glomerular perfusion and filtration during acute extracellular fluid volume depletion. J Clin Invest 75: 1477-1487

Prostanoids and the Renal Response to Hypoxia

W. Jelkmann and C. Weiss

Institute of Physiology, Medical University of Luebeck, Luebeck, FRG

Introduction

There are several specific features of the mammalian kidney with regard to its blood and thus O_2 supply. Under basal conditions, the blood flow per kidney weight is higher and the arteriovenous O_2 difference lower than in any other major organ. Total renal blood flow remains relatively constant over a wide range of arterial blood pressure changes. A decrease in renal blood flow elicits a proportional reduction in renal O_2 consumption (cf. Kiil 1971). Remember, too, the unique architecture of the renal microcirculation (Zimmerhackl et al. 1985). The post-glomerular, second capillary plexus exhibits distinct patterns in different areas of the kidney. Because of the countercurrent flow, the blood passes medullary locations twice. Perhaps as a result of plasma skimming, the concentration of red cells in medullary vessels is much lower than in the systemic circulation.

Thus, despite the high total blood flow and the low arteriovenous difference, the oxygen tension (pO_2) is rather low in medullary areas of the kidney. Note that the pO_2 of mixed venous blood reflects the adequacy of the O_2 supply to the tissue as a whole, but not to the single cells involved. There is significant arteriovenous shunt diffusion of O_2 in the kidney. Microelectrode measurements have revealed a steep pO_2 gradient from the kidney cortex to the medulla (Aperia et al. 1968; Leichtweiss et al. 1969). Epstein et al. (1982) have concluded from spectrophotometric determinations of the redox state of cytochrome aa_3 that substantial portions of the kidney, particularly in the medulla, operate on the brink of anoxia under normal conditions. Even in the kidney cortex hypoxic regions exist (Leichtweiss et al. 1969; Baumgärtl et al. 1972; Franke and Gronow 1986).

In this chapter, the effects of hypoxia on kidney prostanoid production are discussed. Prostanoids are involved in the control of several renal functions, including control of blood flow, renin and kallikrein release, and tubular salt and water reabsorption (Jackson et al. 1985; McGiff and Miller 1986; Dunn 1986). In addition, with regard to mechanisms of O_2 sensing, the capacity of the kidney to release erythropoietin deserves special interest, because the production of this glycoprotein hormone depends mainly on the O_2 supply to demand ratio (cf. Jelkmann 1986). The possible role of prostanoids in this reaction will be dealt with in more detail here.

Prostanoids and Kidney Function

Biochemistry

The principal precursor of prostanoids is arachidonic acid (eicosatetraenoic acid), a 20-carbon polyunsaturated fatty acid, that is cleaved from cellular phospholipids by the interaction of phospholipase A_2 (Fig. 1). The release of arachidonic acid is, in general, the rate-limiting step in prostanoid synthesis. Because the activity of phospholipase A_2 is Ca^{2+}-dependent, prostaglandin synthesis will be stimulated when the cytosolic Ca^{2+} concentration rises either as a result of an increased Ca^{2+} flux across the plasma membrane or the mobilization of Ca^{2+} from intracellular stores. The stimulation of prostanoid synthesis by hormones and neurotransmitters seems to be mediated by phospholipase C initiating the hydrolysis of membrane inositol phospholipids to form inositol 1, 4, 5-trisphosphate and diacylglycerol, which function as intracellular second messengers (Michell 1975; Berridge and Irvine 1984).

Arachidonic acid can be metabolized in one of the following enzymatic ways of oxygenation. The major cyclooxygenase pathway yields the unstable cyclic endoperoxides, PGG_2 and PGH_2, from which the prostaglandins (PGE_2, PGD_2, $PGF_{2\alpha}$), prostacyclin (PGI_2), and the thromboxanes (TxA_2, TxB_2) derive. Alternatively, lipoxygenases initiate the formation of hydroperoxy- and hydroxyeicosa-

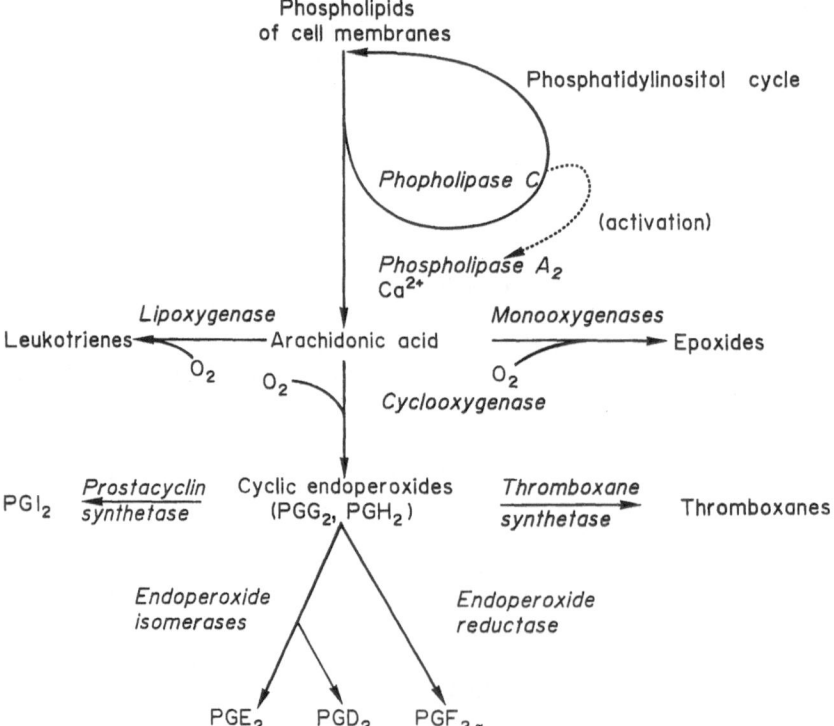

Fig. 1. The formation of eicosanoids from arachidonic acid

tetraenoic acid and leukotrienes. Finally, arachidonic acid may be converted by cytochrome P450-dependent monooxygenases.

More or less, all types of renal tissue have the capacity for producing the aforesaid eicosanoids. However, since only the prostanoid system has been studied with respect to the effects of hypoxia, the reader is referred to the excellent monographs by Jackson et al. (1985), McGiff and Miller (1986), and Dunn (1986) for the functions of the other eicosanoids.

Physiology

Several hormones and neurotransmitters stimulate the production of prostanoids in the kidney, including angiotensin II, kinins, antidiuretic hormone, and catecholamines (Table 1). Cell culture studies have enabled investigators to localize the intrarenal site of action of these agents (Hassid 1982; Morel and Doucet 1986). In addition, the production of the individual prostanoids varies depending both on the kind of stimulus and on the intrarenal site of stimulation. It seems reasonable to assume that the increase in renal prostanoid synthesis occurring under pathophysiological conditions such as ischemia, hydronephrosis, or glomerulonephritis is mediated by the release of polypeptidic autacoids (Table 1).

In turn, the release of prostanoids contributes to the maintenance of normal renal blood flow, glomerular filtration and excretion of sodium chloride and water (Table 2). Note that kidney function in normal subjects is little compromised fol-

Table 1. Stimuli of renal prostanoid synthesis[a]

Hormones and neurotransmitters	Angiotensin II
	Kinins
	Antidiuretic hormone
	Catecholamines
Renal diseases	Impaired blood flow
	Nephritis
	Hydronephrosis

[a] Modified after Dunn and Zambraski (1980)

Table 2. Effects of prostanoids on kidney functions ($+$, stimulation; $-$, inhibition)[a]

	PGE_2 (PGE_1)	PGI_2	Selected references
Renal blood flow (significant redistribution to inner cortex and medulla)	$+$	$+$	McGiff et al. (1974); Bolger et al. (1978)
Sodium chloride reabsorption	$-$	$-$	Kauker (1977); Stokes (1979)
Water reabsorption	$-$		Grantham and Orloff (1968)
Renin release	$+$	$+$	Larsson et al. (1976); Gerber et al. (1978)
Erythropoietin production	$+$	$+$	Gross et al. (1976a); Fisher and Hagiwara (1984); Kurtz et al. (1985)

[a] $PGF_{2\alpha}$ is in general ineffective or has opposite effects compared with PGE_2 and PGI_2

lowing the acute application of prostanoid synthesis-inhibiting drugs such as aspirin or indomethacin. However, the inhibition of prostanoid synthesis may cause deleterious reductions in total and especially inner cortical and medullary renal blood flow, glomerular filtration, and urine production in hemorrhagic or renal insufficient patients (Dunn and Zambraski 1980).

Effects of Hypoxia on Kidney Function and the Role of Prostanoids

Two kinds of O_2-sensitive reactions of kidney cells can be distinguished. Like in other organs, a lowered O_2 supply will elicit changes in blood flow and cellular energetics in the kidney. In addition, however, there is a distinct renal O_2-sensing mechanism that controls the production of the hormone erythropoietin.

General Renal Reactions to Hypoxia

O_2-dependent enzymes, i.e., oxidases and oxygenases, are considered to be the primary sensors of changes of tissue pO_2. Jones (1986a) has pointed out that there are at least 30 such enzymes in the kidney. Their Michaelis constants range from <1 to 300 mmHg O_2. In renal cell suspensions, Weiss (1968) measured a critical pO_2 of 16.5 mmHg below which O_2 consumption decreased. Also, significant intracellular pO_2 gradients are of note (Jones 1986b).

As pointed out above, the local pO_2 differs greatly within normal kidney. Also, if one considers pathophysiological conditions leading to a reduced O_2 supply, it becomes clear, that anemia, reduced renal blood flow, or lowered arterial pO_2 have differential effects on tissue pO_2 in the kidney cortex compared with the medulla. The O_2 supply of the glomerular cells in the kidney cortex is thought to depend mainly on the arterial pO_2 and much less on the blood O_2-carrying capacity or the flow rate (Jelkmann 1986).

When the hematocrit – more exactly, the O_2-transport capacity of blood – is lowered, the O_2 saturation of renal venous blood decreases. Isovolemic anemia elicits an increase in cardiac output. Although renal blood flow was found to increase (cf. Grupp et al. 1972) or to remain constant (Fan et al. 1980), when hematocrit was reduced to 0.15, there is agreement that the renal flow fraction decreases in severe anemia due to local vasoconstriction, preferably in the cortex (Fan et al. 1980). Indeed, Aperia et al. (1968) measured lowered pO_2 values in the kidney cortex of anemic dogs, while the pO_2 increased even above normal in the medulla.

Renal vasodilatory prostanoids are thought to protect the kidney against ischemic effects of catecholamines and angiotensin II during hypotensive hemorrhage. The inhibition of prostanoid synthesis in hemorrhagic dogs resulted in significantly lowered renal blood flow and glomerular filtration (Leffler and Passmore 1977; Henrich et al. 1978; Oliver et al. 1981; Henrich et al. 1981). Needleman et al. (1974) earlier showed that short-term ischemia stimulates prostaglandin production in the isolated perfused rabbit kidney. In addition, Gross et al. (1976b) and Dunn et al. (1978) have reported an increase in serum PGE activity in dogs following constriction of the renal artery (Fig. 2).

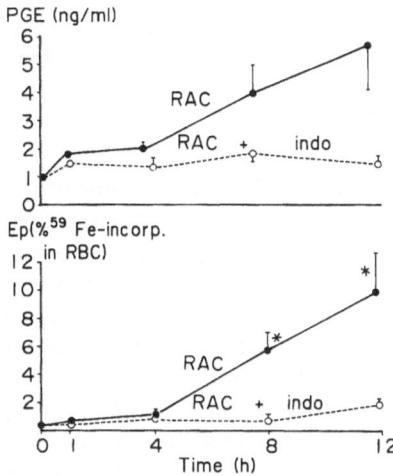

Fig. 2. The increase of serum prostaglandin E *(PGE)* and plasma erythropoietin *(Ep)* following renal artery constriction to 30% of normal blood flow in control dogs *(RAC)* and in indomethacin-treated dogs *(RAC+ indo)*. Redrawn from Gross et al. (1976b)

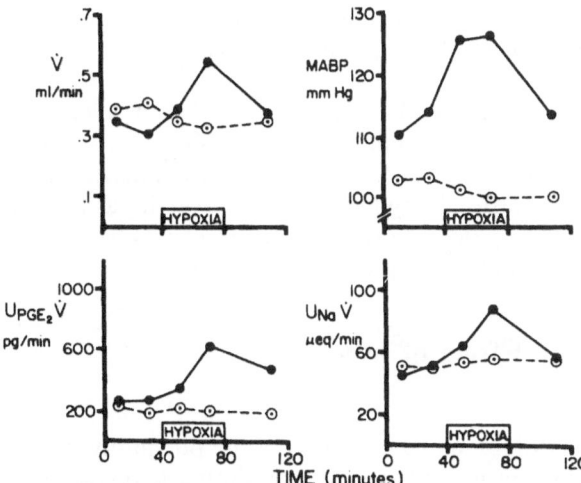

Fig. 3. Urine flow *(V̇)*, mean arterial blood pressure *(MABP)*, PGE_2 excretion *($U_{PGE_2}V̇$)*, and sodium excretion *($U_{Na}V̇$)* in conscious dogs (*n*=6; mean data are shown) exposed to intermittent hypocapnic hypoxia (10% O_2, *closed symbols*) or to air *(open symbols)*. Reproduced from Walker (1982)

Acute hypoxemia results in diuresis and lowered plasma volume. There are controversial reports on the accompanying changes in renal blood flow and glomerular filtration rate (cf. Walker 1982). The discrepancies perhaps resulted from differences in the species under study, the use of anesthetics, and the severity of hypoxia. There is some evidence that the inhibition of prostanoid synthesis during hypoxemia leads to decreased renal blood flow, glomerular filtration, and urine production (Mujovic and Fisher 1975; Weismann 1981). Whereas Weismann (1981) failed to demonstrate an effect of hypoxemia on urinary excretion of PGE and $PGF_{2\alpha}$ in anesthetized dogs, Walker (1982) observed an increase in urinary PGE_2 excretion in conscious dogs during acute hypoxic exposure (Fig. 3). Walker (1982) also found increases in renal blood flow (but unchanged renal vascular resistance), glomerular filtration rate, sodium and potassium excretion, and urine flow during the hypoxemic period. Elevated renal venous plasma titers of prosta-cyclin and thromboxane in hypoxemic dogs have been reported by Burdowski et al. (1980).

Erythropoietin Production

Although extrarenal sites may influence the production of erythropoietin in the kidney, the synthesis of the hormone is thought to depend to some extent on renal cell O_2 tension. Several studies have shown that isolated perfused kidneys and kid-ney cells in culture have the capacity to generate erythropoietin-like activity when maintained at lowered pO_2 (cf. Jelkmann et al. 1986). In addition, the production of erythropoietin increases, when the blood supply to the kidney is attenuated far below normal due to renal artery constriction (Fisher and Samuels 1967).

Initial evidence for a role of prostanoids in the regulation of erythropoietin production during hypoxia has been provided with the finding that the applica-tion of the cyclooxygenase-inhibitor drug indomethacin suppressed the increase in plasma erythropoietin in dogs after renal artery constriction (Mujovic and Fisher 1974; Gross et al. 1976b; Fig.2) or exposure to hypobaric hypoxia (Mujovic and Fisher 1975). Similarly, the inhibition of prostaglandin synthesis was found to re-duce the plasma erythropoietin concentration in hypoxemic (Susić et al. 1979) or anemic (Jelkmann et al. 1984) rats.

The ability of prostaglandins of the E and A series to stimulate in vivo eryth-ropoiesis was first shown by Schooley and Mahlmann (1971). PGE_2, PGA_2 (Gross et al. 1976a) and arachidonic acid (Foley et al. 1978) proved to enhance the re-lease of erythropoietin in isolated perfused dog kidneys. Moreover, it has been re-ported that the production of erythropoietin in renal cell cultures involves prosta-noid-dependent mechanisms (Hagiwara et al. 1984; Kurtz et al. 1985).

Cellular Mechanisms of the Effect of pO_2 on Prostanoid Production

The results of the above studies did not allow exclusion of the possibility that the stimulation of prostanoid production in the intact kidney during hypoxia results solely from an increased activity of humoral agents, such as angiotensin II or cate-cholamines. Therefore, we investigated the effect of pO_2 on in vitro prostaglandin production in renal cell cultures.

In rat kidney mesangial cell cultures, the formation of PGE_2, $PGF_{2\alpha}$ and 6-ke-to-$PGF_{1\alpha}$ – as a measure of prostacyclin synthesis – increased by about the same rate, when the pO_2 in the incubator was lowered from 143 to 14 or 7 mmHg (Jelk-mann et al. 1985; Roszinski and Jelkmann 1987). When the pO_2 was further re-duced to 3 mmHg, prostanoid formation was impaired (Fig.4). Roszinski and Jelkmann (1987) have argued that, at very low pO_2 values – or anoxia – the avail-ability of O_2 in the cyclooxygenase reaction (Fig.1) may become rate-limiting for prostaglandin synthesis. The inhibition of eicosanoid formation in anoxia has been confirmed in bovine endothelial cell cultures (Madden et al. 1986).

Because an enhanced NaCl transport resulted in an increase in PGE_2 synthesis in the canine kidney cell line MDCK, Kurtz et al. (1986) have proposed that the

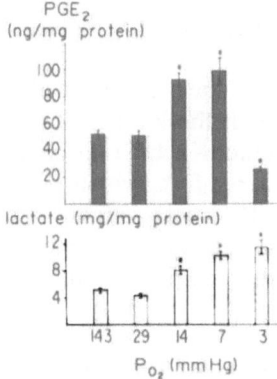

Fig. 4. Effects of pO_2 on PGE_2 and lactate production in rat kidney mesangial cell cultures during 24-h incubation (mean ± SEM; $n = 5$; * significantly different from the value at 143 mmHg O_2). Reproduced from Roszinski and Jelkmann (1987)

production of prostanoids in tubule cells may also be pO_2-dependent. However, in experiments more to the point, Roszinski and Jelkmann (1987) failed to demonstrate an effect of pO_2 changes on PGE_2 production in cultures of the porcine kidney tubule cell lines PK-15 and LLC-PK$_1$. In studying effects of pO_2 on prostanoid production in kidney medulla slices, Zenser et al. (1977) have earlier found that the synthesis of PGE and PGF increased, when the concentration of O_2 in the incubation bath was changed from 5% to 95%.

It would seem, therefore, that mesangiocytes are the only type of renal cell that have clearly proved to produce more prostaglandin when they were maintained at low pO_2 values (7-14 mmHg O_2), but not anoxic. The exact mechanism of stimulation of prostanoid synthesis in these cells remains to be elucidated. Attempts failed to demonstrate an influence of lowered pO_2 on Ca^{2+} influx across the cell membrane. Ca^{2+} augments the activity of phospholipase A_2 (Fig. 1). In fact, the observation of an enhanced Ca^{2+} efflux at pO_2 7 mmHg (Roszinski and Jelkmann 1987) could be indicative of an increased concentration of free Ca^{2+}-ions in the cytosol. Markelonis and Garbus (1975) have earlier considered the possibility that Ca^{2+} is released from mitochondrial matrix during hypoxia. Alternatively, Gunn et al. (1985) have pointed out that a lowered reacylation – rather than an increased release – of arachidonic acid because of a depletion of ATP could be responsible for the stimulation of prostanoid production during hypoxia.

Conclusions

Evidence has been summarized which suggests that the synthesis of prostanoids in the kidney is stimulated by a decrease in the O_2 supply induced by hypoxemia, hemorrhage, or ischemia. In turn, the release of the vasodilatory prostanoids PGE_2 and prostacyclin is thought to play an important role in the maintenance of renal blood flow, especially in medullary areas. It is postulated that prostanoids thereby modulate the vasoconstrictory action of angiotensin II and catecholamines. It is further assumed that PGE_2 and prostacyclin augment the excretion of salt and water, thus contributing to the decrease in blood plasma volume during hypoxemia.

In addition, the effect of hypoxia on prostanoid synthesis seems to be important in the regulation of erythropoietin production.

Apparently, all types of renal cells – more or less – have the capacity for eicosanoid production. The cell type of origin, along with the kind of stimulus, probably determines the pattern of the individual eicosanoids which are derived. Mesangiocytes are the only known type of renal cell, heretofore, which produce prostanoids in cell culture in a pO_2-dependent manner. The exact cellular mechanism of this reaction remains to be clarified. It is postulated that an intracellular movement of Ca^{2+} occurs which leads to the activation of phospholipase A_2, thereby initiating the release of arachidonic acid.

Acknowledgements. The authors wish to acknowledge the expert assistance of Mrs. Gisela Thaler in the typing of this manuscript. Parts of this study were supported by the Deutsche Forschungsgemeinschaft (DFG Grant Je 95/5-1)

References

Aperia AC, Liebow AA, Roberts LE (1968) Renal adaptation to anemia. Circ Res 22: 489-500

Baumgärtl H, Leichtweiss HP, Lübbers DW, Weiss CH, Huland H (1972) The oxygen supply of the dog kidney: measurements of intrarenal pO_2. Microvasc Res 4: 247-257

Berridge MJ, Irvine RF (1984) Inositol trisphosphate, a novel second messenger in cellular signal transduction. Nature 312: 315-321

Bolger PM, Eisner GM, Ramwell PW, Slotkoff LM, Corey EJ (1978) Renal actions of prostacyclin. Nature 271: 467-469

Burdowski AJ, Brookins J, Kadowski PJ, Jubiz W, Salmon J, Moncada S, Fisher JW (1980) Relationships between prostanoids and renin on erythropoietin (Ep) production following hypoxic stimulation. Exp Hematol [Suppl 8]: 310-311

Dunn MJ (1986) The relationship of prostaglandins and thromboxane to the pathophysiology of renal disease and hypertension. In: Fisher JW (ed) Kidney hormones, vol III. Academic, London, pp 397-461

Dunn MJ, Zambraski EJ (1980) Renal effects of drugs that inhibit prostaglandin synthesis. Kidney Int 18: 609-622

Dunn MJ, Liard JF, Dray F (1978) Basal and stimulated rates of renal secretion and excretion of prostaglandins E_2, F_α, and 13, 14-dihydro-15-keto F_α in the dog. Kidney Int 13: 136-143

Epstein FH, Balaban RS, Ross BD (1982) Redox state of cytochrome aa_3 in isolated perfused rat kidney. Am J Physiol 243: F356-F363

Fan F-C, Chen RYZ, Schuessler GB, Chien S (1980) Effects of hematocrit variations on regional hemodynamics and oxygen transport in the dog. Am J Physiol 238: H545-H552

Fisher JW, Hagiwara M (1984) Effects of prostaglandins on erythropoiesis. Blood Cells 10: 241-260

Fisher JW, Samuels AI (1967) Relationship between renal blood flow and erythropoietin production in dogs. Proc Soc Exp Biol Med 125: 482-485

Foley JE, Gross DM, Nelson PK, Fisher JW (1978) The effects of arachidonic acid on erythropoietin production in exhypoxic polycythemic mice and the isolated perfused canine kidney. J Pharmacol Exp Ther 207: 402-409

Franke H, Gronow G (1986) Effects of oxygen tension on kidney cellular energetics. In: Fisher JW (ed) Kidney hormones, vol III. Academic, London, pp 177-216

Gerber JG, Branch RA, Nies AS, Gerkens JF, Shand DG, Hollifield J, Oates JA (1978) Prostaglandins and renin release: II. Assessment of renin secretion following infusion of PGI_2, E_2 and D_2 into the renal artery of anesthetized dogs. Prostaglandins 15: 81-88

Grantham JJ, Orloff J (1968) Effect of prostaglandin E_1 on the permeability response of the isolated collecting tubule to vasopressin, adenosine 3', 5'-monophosphate, and theophylline. J Clin Invest 47: 1154-1161

Gross DM, Brookins J, Fink GD, Fisher JW (1976a) Effects of prostaglandins A_2, E_2 and $F_{2\alpha}$ on erythropoietin production. J Pharmacol Exp Ther 198: 489-496

Gross DM, Mujovic VM, Jubiz W, Fisher JW (1976b) Enhanced erythropoietin and prostaglandin E production in the dog following renal artery constriction. Proc Soc Exp Biol Med 151: 498-501

Grupp I, Grupp G, Holmes JC, Fowler NO (1972) Regional blood flow in anemia. J Appl Physiol 33: 456-461

Gunn MD, Sen A, Chang A, Willerson JT, Buja LM, Chien KR (1985) Mechanisms of accumulation of arachidonic acid in cultured myocardial cells during ATP depletion. Am J Physiol 249: H1188-H1194

Hagiwara M, McNamara DB, Chen I-L, Fisher JW (1984) Role of endogenous prostaglandin E_2 in erythropoietin production and dome formation by human renal carcinoma cells in culture. J Clin Invest 74: 1252-1261

Hassid A (1982) Regulation of prostaglandin biosynthesis in cultured cells. Am J Physiol 243: C205-C211

Henrich WL, Anderson RJ, Berns AS, McDonald KM, Paulsen PJ, Berl T, Schrier RW (1978) The role of renal nerves and prostaglandins in control of renal hemodynamics and plasma renin activity during hypotensive hemorrhage in the dog. J Clin Invest 61: 744-750

Henrich WL, Pettinger WA, Cronin RE (1981) The influence of circulating catecholamines and prostaglandins on canine renal hemodynamics during hemorrhage. Circ Res 48: 424-429

Jackson EK, Branch RA, Margolius HS, Oates JA (1985) Physiological functions of the renal prostaglandin, renin, and kallikrein systems. In: Seldin DW, Giebisch G (eds) The kidney: physiology and pathophysiology. Raven, New York, pp 613-644

Jelkmann W (1986) Renal erythropoietin: properties and production. Rev Physiol Biochem Pharmacol 104: 139-215

Jelkmann W, Kurtz A, Seidl J, Bauer C (1984) Mechanisms of the renal glomerular erythropoietin production. In: Grote J, Witzleb E (eds) Atemgaswechsel und O_2-Versorgung der Organe. Akad Wiss Lit, Mainz, pp 130-137

Jelkmann W, Kurtz A, Förstermann U, Pfeilschifter J, Bauer C (1985) Hypoxia enhances prostaglandin synthesis in renal mesangial cell cultures. Prostaglandins 30: 109-118

Jelkmann W, Kurtz A, Bauer C (1986) In vitro production of erythropoietin. In: Fisher JW (ed) Kidney hormones, vol III. Academic, London, pp 559-583

Jones DP (1986a) Renal metabolism during normoxia, hypoxia, and ischemic injury. Ann Rev Physiol 48: 33-50

Jones DP (1986b) Intracellular diffusion gradients of O_2 and ATP. Am J Physiol 250: C663-C675

Kauker ML (1977) Prostaglandin E_2 effect from the luminal side on renal tubular ^{22}Na efflux: tracer microinjection studies. Proc Soc Exp Biol Med 154: 274-277

Kiil F (1971) Blood flow and oxygen utilization by the kidney. In: Fisher JW (ed) Kidney hormones. Academic, London, pp 1-30

Kurtz A, Jelkmann W, Pfeilschifter J, Bauer C (1985) Role of prostaglandins in hypoxia-stimulated erythropoietin production. Am J Physiol 249: C3-C8

Kurtz A, Pfeilschifter J, Brown CDA, Bauer C (1986) NaCl transport stimulates prostaglandin release in cultured renal epithelial (MDCK) cells. Am J Physiol 250: C676-C681

Larsson C, Weber P, Änggard E (1976) Stimulation and inhibition of renal PG biosynthesis: effects on renal blood flow and on plasma renin activity. Acta Biol Med Ger 35: 1195-1200

Leffler CW, Passmore JC (1977) Effects of indomethacin on hemodynamics of dogs in refractory hemorrhagic shock. J Surg Res 23: 392-399

Leichtweiss H-P, Lübbers DW, Weiss Ch, Baumgärtl H, Reschke W (1969) The oxygen supply of the rat kidney: measurements of intrarenal pO_2. Pflugers Arch 309: 328-349

Madden MC, Vender RL, Friedman M (1986) Effect of hypoxia on prostacyclin production in cultured pulmonary artery endothelium. Prostaglandins 31: 1049-1062

Markelonis G, Garbus J (1975) Alterations of intracellular oxidative metabolism as stimuli evoking prostaglandin biosynthesis. Prostaglandins 10: 1087-1106

McGiff JC, Miller MJS (1986) Renal functional aspects of eicosanoid-dependent mechanisms. In: Fisher JW (ed) Kidney hormones, vol III. Academic, London, pp 363-395

McGiff JC, Crowshaw K, Itskovitz HD (1974) Prostaglandins and renal function. Fed Proc 33: 39-47

Michell RH (1975) Inositol phospholipids and cell surface receptor function. Biochim Biophys Acta 415: 81–147

Morel F, Doucet A (1986) Hormonal control of kidney functions at the cell level. Physiol Rev 66: 377–468

Mujovic VM, Fisher JW (1974) The effects of indomethacin on erythropoietin production in dogs following renal artery constriction. I. The possible role of prostaglandins in the generation of erythropoietin by the kidney. J Pharmacol Exp Ther 191: 575–580

Mujovic VM, Fisher JW (1975) The role of prostaglandins in the production of erythropoietin (ESF) by the kidney. II. Effects of indomethacin on erythropoietin production following hypoxia in dogs. Life Sci 16: 463–473

Needleman P, Minkes MS, Douglas JR (1974) Stimulation of prostaglandin biosynthesis by adenine nucleotides. Circ Res 34: 455–460

Oliver JA, Sciacca RR, Pinto P, Cannon PJ (1981) Participation of the prostaglandins in the control of renal blood flow during acute reduction of cardiac output in the dog. J Clin Invest 67: 229–237

Roszinski S, Jelkmann W (1987) Effect of pO_2 on prostaglandin E_2 production in renal cell cultures. Respir Physiol 70: 131–141

Schooley JC, Mahlmann LJ (1971) Stimulation of erythropoiesis in plethoric mice by prostaglandins and its inhibition by antierythropoietin. Proc Soc Exp Biol Med 138: 523–524

Stokes JB (1979) Effect of prostaglandin E_2 on chloride transport across the rabbit thick ascending limb of Henle. J Clin Invest 64: 495–502

Susić D, Milenković P, Pavlović-Kentera V (1979) The effect of aspirin on erythropoietin formation in the rat. Proc Soc Exp Biol Med 161: 476–478

Walker BR (1982) Diuretic response to acute hypoxia in the conscious dog. Am J Physiol 243: F440–F446

Weismann DN (1981) Altered renal hemodynamic and urinary prostaglandin response to acute hypoxemia after inhibition of prostaglandin synthesis in the anesthetized dog. Circ Res 48: 632–640

Weiss Ch (1968) Critical oxygen tension and rate of respiration of isolated kidney cells. In: Lübbers DW, Luft UC, Thews G, Witzleb E (eds) Oxygen transport in blood and tissue. Thieme, Stuttgart, pp 227–237

Zenser TV, Levitt MJ, Davis BB (1977) Effect of oxygen and solute on PGE and PGF production by rat kidney slices. Prostaglandins 13: 143–151

Zimmerhackl B, Robertson CR, Jamison RL (1985) The microcirculation of the renal medulla. Circ Res 57: 657–667

Oxygen Tension and Erythropoietin Production: The Role of the Macrophage in Regulating Erythropoiesis

I. N. Rich

Department of Transfusion Medicine, University of Ulm, and the German Red Cross Blood Bank, Ulm, FRG

Introduction

It is not known at what stage during ontogeny erythropoietin (Epo) first becomes important in the production of red blood cells. Thus, it is unknown whether Epo is a prerequisite for yolk sac erythropoiesis or, for that matter, erythropoiesis in any of the other organs, i.e., fetal liver, spleen, and bone marrow. Nor is it known whether the reason for support of erythropoiesis in these organs during ontogeny (and erythropoietic stimulation whereby erythropoiesis retraces its path of ontogeny) is due to the production of Epo by specific cells present in these organs. The latter notion may be supported by the fact that in the fetus, the liver is the sole site of Epo production (Zanjani et al. 1977) and the fetal liver macrophage the cellular origin (Gruber et al. 1977).

At some time during the later stages of fetal erythropoiesis, before, during, or after the transition to the bone marrow, Epo production has been considered to be taken over by the kidney. However, neither the anatomical site nor the cellular location within the kidney is yet known, although at least two possibilities have been proposed, namely, the tubular cell (Caro et al. 1987) and the mesangial cell (Bauer 1987). To date, there is little convincing evidence that the kidney is actually responsible for the day-to-day production of erythropoietin, although there is sufficient information, particularly at the molecular level, showing that Epo is produced under hypoxic conditions (Bondurant and Koury 1986; Goldwasser et al. 1987; Caro et al. 1987). To some readers, the first part of this latter statement may appear heresy, since it has been assumed that the kidney is the primary Epo production site in the adult mammal. However, it is safe to conclude that at present, at a molecular level, no messenger RNA for Epo has been detected by Northern transfer from either whole kidney or kidney section extracts (Bondurant and Koury 1986; Goldwasser et al. 1987; Caro et al. 1987). It has therefore become imperative to differentiate between the normal, steady-state, day-to-day production of Epo and that which occurs under perturbated or pathophysiological conditions.

Erythropoietin production is increased under hypoxic and decreased under polycythemic conditions. This in turn is dependent on the oxygen-carrying capacity of the circulating erythrocyte population, the mass of which is in turn, absolutely dependent on the Epo concentration and therefore the prevailing oxygen tension. We do not know how the oxygen availability is sensed by specific cells in order to regulate Epo production and therefore erythropoiesis; that is, we do not know the cellular and molecular mechanism whereby molecular or diffused oxygen, appearing on the cell membrane is sensed as a signal which is transferred into the cell causing the eventual modulation of hormone production and/or release.

What then do we know about oxygen sensing and Epo production? The answer is very simple. Precious little! Biological oxygen sensing in general is, to coin a well-used phrase, a black box. We know that aerobic organisms live in a state of irony; the oxygen that is an absolute necessity for life can also be a death sentence due to oxygen toxicity caused by the production of free oxygen radicals. To this end, animal cells and in particular mammalian cells have evolved complex systems in order to combat or reduce oxygen toxicity. Is it possible that one or more of these systems are receptors for oxygen tension which in some manner passes the information to the genetic, translation, glycosylation (if necessary), or secretory machinery of the cell? Or is oxygen sensing the result of the overall cellular activity that in some manner is biochemically detected by components in cellular organelles such as the mitochondria (Jobis 1977)?

The following discussion reviews some of the results we have obtained with an in vitro model for Epo production under physiological oxygen tension and the efforts we are making to try to extrapolate them to the in vivo situation. It should be emphasized at the outset that, like many other investigations in the field of oxygen sensing, the results primarily describe the effects of oxygen tension on cellular functions and give little or no information as to the mechanism underlying these effects.

Oxygen Tension and Hemopoiesis

The effects of oxygen tension on intact biological systems, in vivo and in vitro models are too numerous to mention here. It is safe to say, however, that excluding reports in which hypoxia has been used as a signal for erythropoietic stimulation, publications relating to the direct effect of oxygen tension on hemopoietic cells are very few in number.

The first experiments involving the effect of low oxygen tensions on the growth of hemopoietic progenitor cells were reported in 1978 by Bradley and coworkers. They showed, using an in vitro colony-forming assay in agar, that myeloid progenitor cells, granulocyte-macrophage colony-forming cells or GM-CFC, were stimulated by granulocyte-macrophage colony-stimulating factor (GM-CSF) to a far greater extent under 7.5% oxygen than under atmospheric oxygen conditions. These experiments were performed in a "closed system" in which the cultures were gassed with a performed gas mixture and sealed in plastic boxes.

The advantageous effect of low oxygen tension on erythropoietic cell growth was reported by Rich and Kubanek in 1982. This in vitro culture system utilized the fact that when erythroid cells from either fetal liver (Axelrad et al. 1974) or adult mouse bone marrow (Iscove and Sieber 1975) were cultured in the presence of erythropoietin in semisolid medium (plasma clot or methyl cellulose), colonies of erythroid cells could be obtained after various periods of time. Those cells giving rise to small colonies (< 64 cells) within 48 h were designated the colony-forming unit-erythroid or CFU-E. This population of cells can be considered to be approximately equivalent to the earliest morphologically identifiable erythroblast, the pronomoblast. Those cells giving rise to macroscopic colonies after about 10 days in culture were designated burst-forming unit-erythroid or BFU-E. This latter population of cells represent the precursors of the CFU-E population and are the direct descendants of the pluripotential hemopoietic stem cells. It was found that by adding thiol-containing compounds to the cultures, an enhancement in colony formation from both CFU-E and BFU-E could be obtained (Iscove and Sieber 1975).

We reasoned that the addition of thiol-containing compounds could result in the reduction of disulfide bridges in certain cellular components and that the addition of naturally occurring thiol compounds involved in reducing oxygen toxicity may also produce the same effect. Indeed, addition of reduced glutathione or vitamin E to the cultures at physiological concentrations resulted in exactly the same effect as the addition of alpha-thioglycerol or beta-mercaptoethanol at 0.1 μM (Rich and Kubanek 1982a). The stage was therefore set to investigate whether reducing the oxygen gas phase also caused a further potentiation in colony formation of erythropoietic cells. By reducing the oxygen present in the incubators from normoxic to 5% oxygen tension conditions, not only was an enhancement in colony formation obtained, but an increase in erythropoietin sensitivity of the cells (Rich and Kubanek 1982a). These results clearly demonstrated that by incorporating natural thiol compounds (e. g., reduced glutathione) together with reduction in the oxygen concentration, oxygen toxicity in the form of free radical production could be dramatically reduced, leading to an increased cell survival and therefore a dramatic enhancement in the plating efficiency of cultured erythropoietic cells (Rich and Kubanek 1982a).

Although 5% oxygen corresponds to the venous partial oxygen tension, it would be expected that due to oxygen gradients (Duling and Berne 1970), the concentration reaching the erythropoietic target cells in the bone marrow would be even further reduced. Experiments were then initiated using 5%, 3.5%, and 2% oxygen concentrations. It was found that for both CFU-E and BFU-E populations, the optimal oxygen concentration was 3.5%, followed by 2% and 5% (Rich 1986b).

The advantageous effects of oxygen tension on hemopoietic cell growth, and erythropoietic cell growth in particular, are slowly becoming adopted in culture protocols (Eliason and Odartchenko 1985; Pennathar-Das and Levitt 1985; Smith and Broxmeyer 1986). By culturing such cells under physiologically relevant oxygen tensions, it is possible to better simulate the physiochemical in vivo environment, even though the original cellular microenvironment has been destroyed by cell preparation techniques.

The Macrophage as an Erythropoietin-Producing Cell

In 1980, we demonstrated that the macrophage-specific, cytotoxic agent crystalline silica (silicon dioxide, SiO_2) could be used to release an erythropoietin-stimulating factor (ESF) into the surrounding medium of macrophage-containing mouse cell suspensions (Rich et al. 1980). The reaction reached maximum release of Epo within 30 min of bringing cells and silica together, a time period too short to destroy the cells (Davies and Allison 1974). Using the erythroid colony-forming technique (CFU-E assay), it was possible to show that addition of the supernatants could increase the number of colonies above the background. The activities present in supernatants from various organs occurred in the following order: spleen > lung > bone marrow > peritoneium > fetal liver > adult liver > kidney (Rich et al. 1980). Later it was possible to show that the ESF released was actually Epo using an in vivo bioassay for the hormone and neutralization with an antiserum (Rich et al. 1981; Rich 1986a).

Because macrophages are particularly difficult entities to study in vivo, we developed a culture system whereby unstimulated and unseparated bone marrow cells were cultured on hydrophobic fluorethylenpropylene (FEP) membranes (Rich and Kubanek 1982). The cells were cultured in suspension either under low serum (5%) or serum-free conditions (Rich 1986a). After 14 days in culture, an almost pure population of macrophages (98%) could be ascertained using both nonspecific esterase staining and a monoclonal antibody directed against the mouse, macrophage-specific F4/80 antigen (Austyn and Gordon 1986). With the increase in macrophages, there was also a concomitant increase in three hemopoietic regulator molecules present in the supernatants, these being Epo, granulocyte-macrophage colony-stimulating factor (GM-CSF) and interleukin-3 (IL-3) or multi-CSF (Rich 1986a).

Thus, the macrophages present in culture were capable of releasing (and presumably also producing) hemopoietic regulator molecules. These results implied that the macrophage, in addition to the vast array of other secretory products, could possibly function as a "regulator cell" at the sites of hemopoieis (Rich 1986a). In fact, functional and phenotypic characterization of the cultured macrophage population demonstrates that the cells are "resident" in nature and not activated or stimulated. The possibility therefore exists that if macrophages in vivo are capable of producing Epo, then they may also be stromal "resident" cells and not those responsible for the inflammatory response.

The Macrophage as an Oxygen-Sensing Cell

The first correlation between changing oxygen tensions and Epo release by the macrophage was found when mice were subjected to either hypertransfusion or bleeding followed by removal of the spleen and bone marrow and treatment of the cell suspensions with crystalline silica to release intracellular Epo (then denoted ESF) into the surrounding medium (Rich et al. 1980). Increasing concentrations of supernatant were added to the in vitro assay employing the stimulation of mouse fetal liver CFU-E to form erythroid colonies after 48 h in culture. Compared to silica-treated bone marrow and spleen cell suspensions from normal mice, which demonstrated a dose-dependent increase in erythroid colony formation with increasing supernatant volumes, no such dose dependency was observed for cells obtained from hypertransfused mice. More important, the number of erythroid colonies obtained from spleen and bone marrow cell suspensions were not significantly different from the background or spontaneous colony formation. In contrast, however, bleeding resulted in an increase of about 2.5-fold for bone marrow and about threefold for spleen above that obtained for unstimulated bone marrow and spleen, respectively. In short, hypertransfusion resulted in a considerable decrease, and bleeding an enhancement in Epo release from macrophages.

The above results were obtained from assays in which the target cells were incubated under normal atmospheric oxygen tension conditions. By culturing unseparated and unstimulated bone marrow cells under 2%, 3.5%, or 5% oxygen concentrations for 14 days and measuring the activities of all three hemopoietic regulator molecules present in the supernatants, the following results were obtained. The optimal oxygen tension by which Epo was released at 14 days was 3.5%, followed by 2% and 5%. The activity of GM-CSF release was not affected by changing oxygen tensions. However, the activity of IL-3/multi-CSF increased with increasing oxygen tension, i.e., 2% < 3.5% < 5%. At no time were either Epo or IL-3/multi-CSF not produced. Several conclusions can be drawn from these results. First, the macrophage is capable of sensing and responding to physiological changes in oxygen tension; that is, the macrophage must possess an oxygen sensor. Second, oxygen sensing and Epo production/release must occur in the same cell since we are dealing with an almost pure population of macrophages. Third, there is no on-off switch controlling these factors, but rather a modulation of activity. This is to be expected if the production of hemopoietic cells is to be continuous throughout the life of the animal (Rich 1986b). Fourth, these oxygen changes are sufficient to cause changes in the release of both, Epo and IL-3/multi-CSF.

Recent experiments have demonstrated that when macrophages derived from 14-day cultures incubated under either of the aforementioned oxygen tensions are added to assays of target cells and cultured under the same oxygen tensions as the macrophages but in the absence of exogenous stimulator, i.e., no standard Epo or supernatants, stimulation of colony formation occurred and was again dependent on the prevailing oxygen tension; the optimal being again 3.5% O_2 (Rich 1987).

The pathway from sensing a change in oxygen tension to the appearance of an increased or decreased number of end cells may be considered to incorporate at least two signal systems. The first signal is oxygen tension, which, by a mechanism we do not yet understand, results in a modulation of Epo production. The second signal is the concentration of Epo itself which upon release binds to the receptor(s) of the target cell(s) and initiates a program calling either for a decreased or increased number of red blood cells. The first signal system would be expected to occur rapidly (within hours), the second much more slowly (days). Preliminary results appear to indicate that indeed the first signal system takes place rapidly. After culturing the macrophages for 14 days in either 2%, 3.5%, or 5% oxygen, the supernatants were removed and replaced with fresh medium. The primary 2% cultures were then incubated under 2%, 3.5%, or 5% oxygen, the 3.5% cultures under 2%, 3.5%, or 5% oxygen and the 5% cultures under 2%, 3.5%, or 5% oxygen. Under these conditions it would be expected that those cultures originally incubated under 2% oxygen tension and transferred to 5% would receive a signal to decrease Epo release and/or production. In contrast, cultures originally incubated under 5% oxygen tension and transferred to 2% oxygen would receive an increased hypoxic stimulus and respond by increasing Epo release and/or production. And this is exactly what happens. More important, however, is that the time taken to transfer the signal of oxygen tension change to a new level of Epo release occurs within 24 h. Experiments are now underway in order to try to determine the earliest possible time at which a change in Epo release is due to a change in physiologically relevant oxygen tensions. In this way, the kinetics of the system can be ascertained and used to look for system that can respond accordingly.

It should be emphasized that the above experiments are not the only ones that indicate that the macrophage can sense and respond to changing oxygen tensions. Macrophages are also capable of producing an angiogenesis factor which stimulates vascularization at the site of wound healing where the pO_2 is about 15 mm Hg or about 2%. It has been shown by Knighton et al. (1983) that production of this factor occurs transiently within a definite oxygen tension range. When the pO_2 is increased above 15 mm Hg, production of the factor ceases. The interesting difference between this system and Epo production is that it would be expected that the macrophage capable of producing angiogenesis factor be "activated" and not "resident" in nature. This would in turn imply that all macrophages are capable of sensing changes in physiological oxygen tensions, but their functional capabilities are different.

The Future

It may be apparent that understanding the cellular and molecular mechanisms involved in oxygen sensing will probably be a situation in which negative results will provide more answers than positive ones; in other words, a process of elimination is necessary. With no direct "handle" on the problem, this may be the only possibility. Indeed, even at a molecular level, a process of elimination seems to be the only alternative.

In a recent workshop on erythropoietin and erythropoietic regulation, considerable time was spent on discussing the regulation of Epo production (Rich 1987). At the molecular level, one interesting possibility proposed was that changing oxygen tensions resulted in a change in the "degree of genetic expression." Of course the pathway linking the appearance of the external signal on the cell membrane to the genetic machinery is another problem that has been discussed for Epo production by kidney cells (Bauer 1987). However, in this case it is envisaged that the problem be tackled not at the cell surface where the signal is received, but at the genetic level where the signal is possibly transformed into a response.

In order to address this problem, we have initiated studies in which a possible genetic mechanism to oxygen sensing could be investigated. Macrophages grown under 2%, 3.5%, or 5% oxygen tension were removed from the culture and cytocentrifuged onto glass slides. In situ hybridization was then performed using a 1.2 kb mouse Epo DNA probe labeled with either 35-S or biotin. Hybridization was detected using either autoradiography or streptavidin conjugated to alkaline phosphatase, respectively. Using both these hybridization procedures and counting the number of positive cells compared with controls, no significant difference was observed between the three different oxygen tensions.

Taken at face value, these results would not confirm that changing oxygen tensions affect the degree of expression of the gene. However, there is still hope. Although no apparent change in the number of positive cells could be correlated with changing oxygen tension, the number of Epo mRNA molecules per cell may be changed. Detection of the hybridization signal using the above two methods is not sufficiently sensitive to answer this question, at least when using cultured macrophages. However, using a biotin-labeled DNA probe, it is possible to employ different signal detection systems.

One such system involves the use of colloidal gold. Recent experiments in our laboratory have shown that a high and low level of hybridization can occur when a biotin-labeled Epo DNA probe and streptavidin-gold detection system coupled with observation under the reflection-contrast microscope are used. Using this in situ hybridization system and observation under the electron microscope, it may be possible to quantitate the hybridization signal and therefore Epo gene expression on a per cell basis when macrophages are cultured under different physiological oxygen tensions. The possibility also exists of extrapolating this system to the in vivo situation. Indeed, it has been possible to demonstrate using the aforementioned in situ hybridization system, that 3% of the normal mouse bone marrow macrophages are capable of expressing the Epo gene. Furthermore, these macrophages appear to be associated intimately with other cells in the form of "blood islands". By subjecting the animals to various degrees of erythropoietin perturbation and examining the erythroid organs for in situ hybridization, it may be possible to examine oxygen sensing at the intact in vivo level. However, our knowledge of the mechanism of oxygen sensing by cells may have to wait until we know more about the molecular and biochemical interactions that can occur when "relatively" small changes of pO_2 take place.

References

Austyn JM, Gordon S (1981) F4/80, a monoclonal antibody directed specifically against the mouse macrophage. Eur J Immunol 11: 805–815

Axelrad AA, McLeod DL, Shreeve MM, Heath DS (1974) Properties of cells that produce erythropoietic colonies in vitro. In: Robinson WA (ed) Hemopoiesis in culture. US Government Printing Office, Washington, pp 266–284

Bauer C (1987) Chemoreception of oxygen in the kidney and erythropoietin production. In: Rich IN (ed) Molecular and cellular aspects of erythropoietin and erythropoiesis. Springer, Berlin Heidelberg New York, pp 311–327

Bondurant MC, Koury MJ (1986) Anemia induces accumulation of erythropoietin mRNA in the kidney and liver. Mol Cell Biol 6: 2731–2733

Bradley TR, Hodgson GS, Rosendaal M (1978) Effect of oxygen tension of hemopoietic and fibroblast cell proliferation in vitro. J Cell Physiol 97: 517–522

Caro J, Schuster S, Besarab A, Erslev AJ (1987) Renal biogenesis of erythropoietin. In: Rich IN (ed) Molecular and cellular aspects of erythropoietin and erythropoiesis. Springer, Berlin Heidelberg New York, pp 329–336

Davies P, Allison AC (1974) Secretion of macrophage enzymes in relation to the pathogenesis of chronic inflammation. In: Nelson DS (ed) Immunology of the macrophage. Excerpta Medica, Amsterdam, pp 427–461

Duling BR, Berne RM (1970) Longitudinal gradients in periarteriolar oxygen tensions: a possible mechanism for the participation of oxygen in local regulation of blood flow. Circ Res 27: 669–678

Eliason JF, Odartchenko N (1985) Colony formation by primitive hemopoietic progenitor cells in serum-free medium. Proc Natl Acad Sci USA 82: 775–779

Goldwasser E, McDonald J, Beru N (1987) The molecular biology of erythropoietin and the expression of its gene. In: Rich IN (ed) Molecular and cellular aspects of erythropoietin and erythropoiesis. Springer, Berlin Heidelberg New York, pp 11–21

Gruber DF, Zucali JR, Mirand EA (1977) Identification of erythropoietin producing cells in fetal mouse liver cultures. Exp Hematol 5: 392–398

Iscove NN, Sieber F (1975) Erythroid progenitors in mouse bone marrow detected by macroscopic colony formation in culture. Exp Hematol 3: 32–43

Jobis FF (1977) What is a molecular oxygen sensor? What is a transduction process? Adv Exp Med Biol 78: 3–18

Knighton DR, Hunt TK, Scheuenstahl H, Halliday BJ, Werb Z, Bonda MJ (1983) Oxygen tension regulates the expression of angiogenesis factor by macrophages. Science 221: 1283–1285

Pennathar-Das R, Levitt L (1985) Hypoxic-induced augmentation of human marrow erythropoiesis by monocytes (MO) and T-cells (abstract). Exp Hematol 13: 381

Rich IN (1986a) A role for the macrophage in normal hemopoiesis. I. Functional capacity of bone marrow derived macrophages to release hemopoietic growth factors. Exp Hematol 14: 738–754

Rich IN (1986b) A role for the macrophage in normal hemopoiesis. II. Effect of varying physiological oxygen tensions on the release of hemopoietic growth factors from bone marrow derived macrophages in vitro. Exp Hematol 14: 746–751

Rich IN (1987) Erythropoietin production by macrophages: cellular response to physiological oxygen tensions and detection of erythropoietin gene expression by in situ hybridization. In: Rich IN (ed) Molecular and cellular aspects of erythropoietin and erythropoiesis. Springer, Berlin Heidelberg New York, pp 291–310

Rich IN, Kubanek B (1982a) The effect of reduced oxygen tension on colony formation of erythropoietic cells in vitro. Br J Haematol 52: 579–588

Rich IN, Kubanek B (1982b) Extrarenal erythropoietin production by macrophages. Blood 60: 1007–1018

Rich IN, Heit W, Kubanek B (1980) An erythropoietic stimulating factor similar to erythropoietin released by macrophages after treatment with silica. Blut 40: 297–303

Rich IN, Anselstetter V, Heit W, Zanjani E, Kubanek B (1981) Release of erythropoietin from macrophages by treatment with silica. J Supramol Struct Cell Biochem 15: 169–176

Smith S, Broxmeyer HE (1986) The influence of oxygen tension on the long-term growth in vitro of haematopoietic progenitor cells from human cord blood. Br J Haematol 63: 29–35

Zanjani ED, Poster J, Burlington H, Mann LI (1977) Liver as the primary site of erythropoietin formation in the fetus. J Lab Clin Med 89 640–644

Effect of Hypoxia on Ca^{2+} Influx and Catecholamine Synthesis in Chemosensitive Cells of the Carotid Body in Tissue Culture

F. Pietruschka

Max-Planck-Institut für Systemphysiologie, Rheinlanddamm 201, 4600 Dortmund 1, FRG

Isolated cells growing in tissue culture in vitro are a good tool to study cell-physiological questions under standardized conditions. This is possible when the cultured cells retain or regain the morphological and biochemical characteristics they possessed in vivo. In the case of carotid body cells one of these typical characteristics would be that the chromaffin type-I cells are enveloped by elongated processes of the type-II cells (sustentacular cells). A short-term culture of carotid body cells shows mainly epithelial cells which extend some processes (Fig. 1). They can be identified as type-I cells by their catecholamine content and their fine structure (Fig. 2). The primary culture also contains small elongated cells, whose electron microscopy is very similar to that of sustentacular cells in vivo (Fig. 3). The catecholamine fluorescence typical for type-I cells increases after 1 week of culturing and is found even in 4-week-old cultures (Pietruschka 1974). Further-

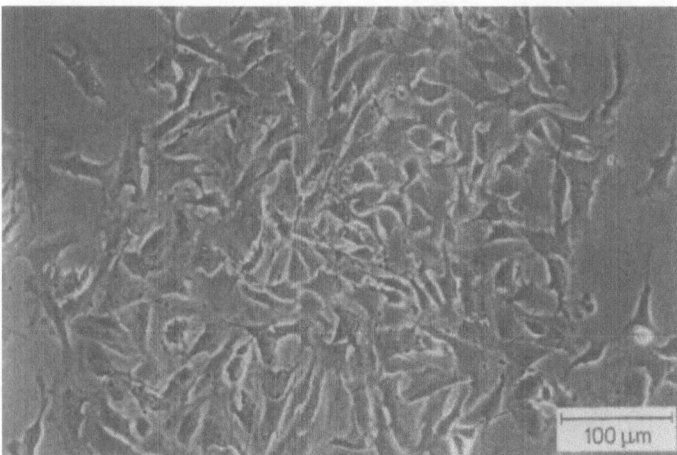

Fig. 1. A 2-day-old culture of carotid body cells shows mainly epithelial cells which extend some processes. Carotid body cells were prepared from rabbit embryos. The cells were dissociated by trypsin/collagenase and plated in multiwell plastic culture dishes in Dulbecco's minimum essential medium + 10% fetal calf serum at 37° C and 10% CO_2 in air. The number of cells varied from about 1×10^4 to 5×10^4 per culture dish

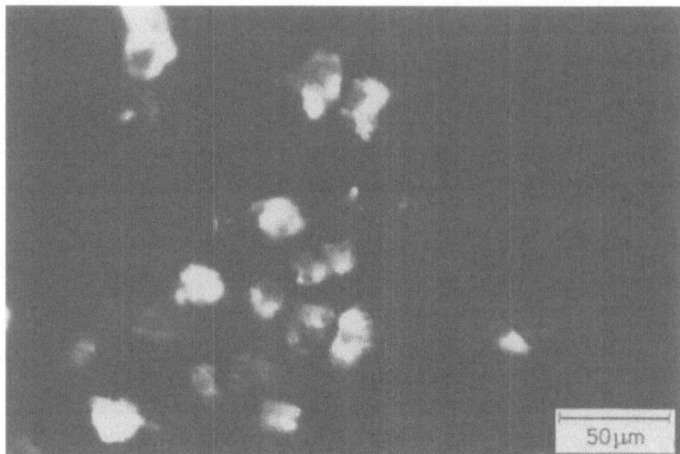

Fig. 2. Cluster of epithelial cells of glomera derived from 11-day-old rabbits after 18 days in culture. For fluorescence microscopy the cultures were briefly rinsed in warm distilled water and immediately frozen in Frigen cooled in liquid nitrogen. They were then freeze-dried and treated with formaldehyde gas for 45 min at 80° C, the paraformaldehyde being equilibrated with air of 60% humidity (Falck et al. 1962)

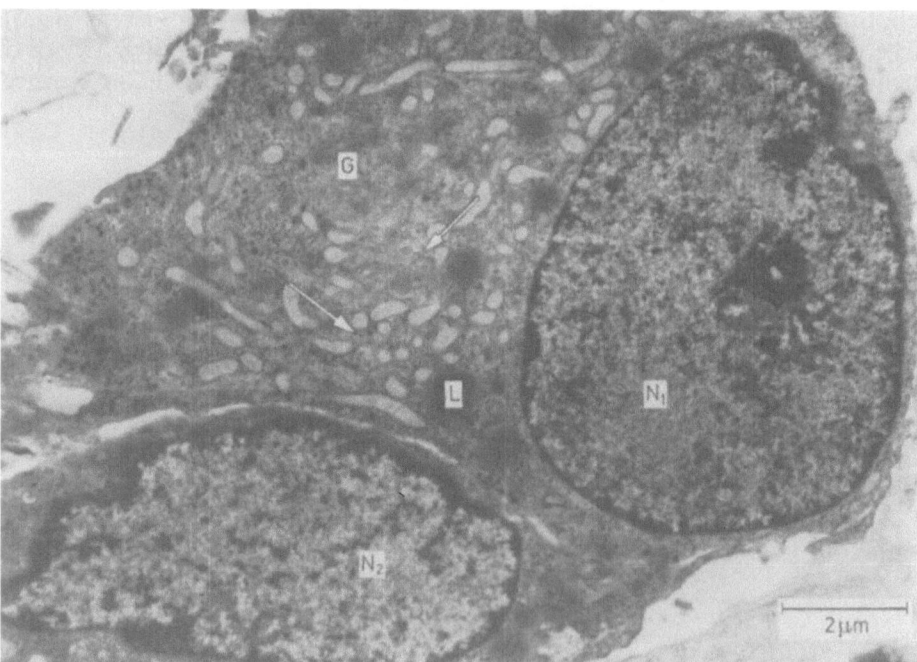

Fig. 3. Cells after 3 days of culturing. *N1*, nucleus of type-I cell; *N2*, nucleus of type-II cell; *G*, Golgi apparatus; *L*, lipid droplet; *arrow*, dense-cored vesicles. For electron microscopic examination the cells were fixed in 0.2 mol/l glutaraldehyde in 0.1 mol/l cacodylate buffer + 0.2 mol/l saccharose (480 mosmol) for 15 min at 4° C

more, the fine structure of both cell types is similar to that in vivo (Fig. 3). Type-I cells show a compact shape, round or oval nuclei, many mitochondria, rough endoplasmatic reticulum, numerous free ribosomes, a well-developed Golgi apparatus, and a great number of dense-cored vesicles. The electron-dense material in the vesicles of dictyosomes and in the granular vesicles near the Golgi apparatus indicates that the synthesis of catecholamines continues in culture (Pietruschka and Schäfer 1976). Type-II cells still envelop the compact glomus cells. Their nuclei are oval or reniform and their cytoplasm has fewer organelles and no dense-cored vesicles.

Experiments with varying gas conditions have shown that a reduction in O_2 concentration to 5% and a rise in CO_2 to 10% stimulates cell differentiation after 3 days of culturing (Pietruschka et al. 1977).

As the carotid body is a chemoreceptor, which is sensitive to hypoxia, it was interesting to see in which way a reduction of O_2 concentration influences the isolated cells. As the first step of activation, we measured the calcium influx into the cells, since calcium is considered to be a second messenger that plays a role in stimulus secretion coupling in endocrine cells (Rasmussen and Barrett 1984). Measurement of the calcium influx in carotid body cells should show whether hypoxia has a direct effect on the cells. As sustentacular cells are similar to Schwann cells and belong to glial elements, we carried out our experiments with glioma cells cultured under the same conditions. This should enable us to determine whether the reaction of carotid body cells to the given stimuli is a specific one, and may also provide information as to which cell type may be responsible for it.

We prepared carotid body cells from rabbit embryos and cultured them for merely 2 days to minimize growth of nonspecific cells. During this time, no mitosis of sustentacular cells or fibroblasts was observed. Glioma cells were derived from the rat glioma cell line, C_6. They were cultivated at similar densities and under the same conditions as carotid body cells (Fig. 4).

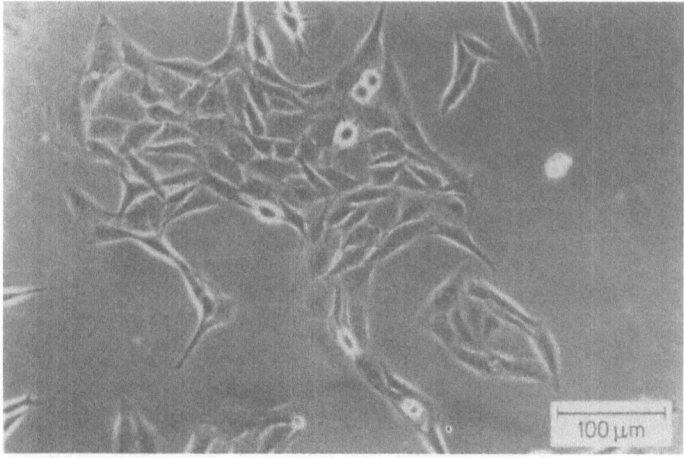

Fig. 4. Glioma cells derived from the rat glioma cell line C_6. They were cultivated at similar densities and under the same conditions as carotid body cells.

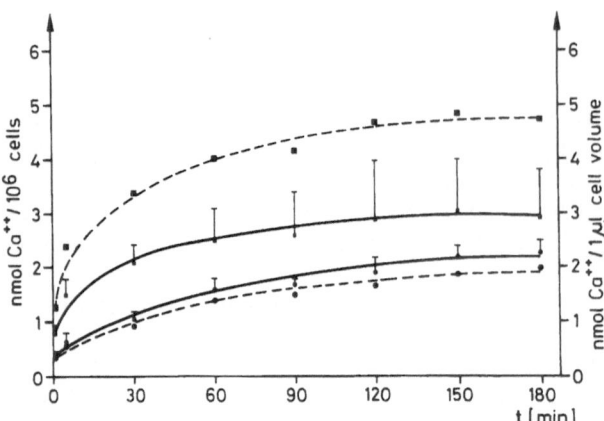

Fig. 5. Ca^{2+} uptake under steady-state conditions into carotid body *(squares)* and glioma cells *(circles)*. *Continuous lines* show Ca^{2+} uptake per cell number (means ± SEM); *dotted lines* show Ca^{2+} uptake based on cell volume. ^{45}Ca^{2+} uptake experiments were performed as described by Barnes and Mandel (1981). Cells were washed and preincubated in HEPES-buffered saline (in mmol/l NaCl 136, KCl 5.4, MgCl$_2$ 1.4, CaCl$_2$ 1.2, NaH$_2$PO$_4$ 1.0, glucose 10, HEPES 20; pH 7.4). After 1 h of preincubation the medium was changed to an incubation medium containing ^{45}CaCl$_2$ (0.8 μCi in 200 μl; New England Nuclear, Boston, Mass.). Uptake experiments lastet from 1 min to 3 h. They were terminated by cooling the culture dishes on an ice-cooled metal plate and four rapid washes in ice-cold solution. Digestion of the cells was performed with 200 μl of 0.5 N NaOH. The ^{45}Ca^{2+} content of the samples was determined in a scintillation counter (Beckmann, Fullerton, Calif.). For comparison with the literature, all values of calcium exchange were calculated for 1×10^6 cells

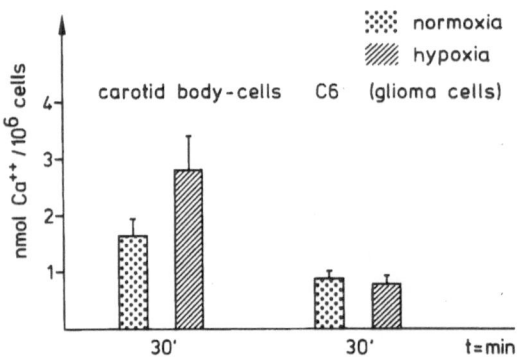

Fig. 6. Effect of hypoxia on calcium uptake. Hypoxia was set up in an atmos bag, an inflatable polyethylene chamber, filled with N$_2$ or air (for controls). After preincubation, culture dishes were transferred to an exsiccator in the bag. When the bag was filled with gas, the dishes were attached to a shaker and the medium was changed to the labeled incubation medium

The exchangeable pool of calcium was measured in carotid body and glioma cells under steady-state conditions from 1 min to 3 h (Pietruschka 1985). Figure 5 shows that the exchange is more rapid in carotid body cells and that the amount of exchangeable calcium is also higher. In carotid body cells we find values of 3 nmol ± 0.86/10^6 cells, compared to 2.3 nmol ± 0.23/10^6 in glioma cells. Based on the cell volume this difference is even greater (4.8 nmol/10^6 cells compared with 1.8 nmol/10^6 cells). The pool of exchangeable calcium in carotid body cells exceeds more than twofold that of glioma cells (Fig. 5).

Supposing that the carotid body cells are chemoreceptors for hypoxia, then hypoxia should cause a rise in calcium uptake in these cells but not in glioma cells. This was indeed the case. In carotid body cells there is a marked stimulation in calcium exchange during a 30-min incubation in hypoxic medium. Calcium exchange increased significantly ($P<0.02$) to 170% of control values (Fig. 6). Rat glioma cells treated in the same way as carotid body cells did not show a comparable difference in Ca^{2+} turnover under hypoxic and normoxic conditions.

In the chromaffin cells of the adrenal medulla, Haycock et al. (1982) have shown that an increase in intracellular calcium concentrations causes a stimulation of the activity of tyrosine hydroxylase. As this enzyme is rate limiting in catecholamine synthesis, a raised catecholamine concentration in carotid body cells during hypoxia could indicate an increase in intracellular calcium. Thus, we compared the response of carotid body cells and neuroblastoma cells (Fig. 7), the latter also show tyrosine hydroxylase activity, but would not be expected to react to hypoxia.

The synthesis of catecholamines was measured in 3-day-old cultures after incubation in ^{14}C-tyrosine under normoxic or hypoxic conditions. The catecholamines were separated by thin-layer chromatography (Fleming and Clark 1979) and the values for the precursors and degradation products of dopamine were determined. These are of interest, since carotid bodies of the rabbit secrete dopamine under hypoxia. About 80% of the radioactivity could be identified by reference standards, most of the label was present as tyrosine. In carotid body cells, only $1.8\% \pm 0.4\%$ of the tyrosine taken up by the cells was used for catecholamine synthesis. This rate was about twice as high as in neuroblastoma.

After a hypoxic stimulation of 5 min followed by 85 min of further incubation in ^{14}C-tyrosine under normoxia (Fig. 8), the concentration of radioactive catecholamines in the carotid body cells rose from $17.5 \ \text{pmol} \pm 7.0/10^6$ cells in controls to $44.6 \ \text{pmol} \pm 17.9/10^6$ cells ($P<0.05$). This increase of 255% was mainly due to DOPA and dopamine formation (Table 1). In neuroblastoma cells there was no

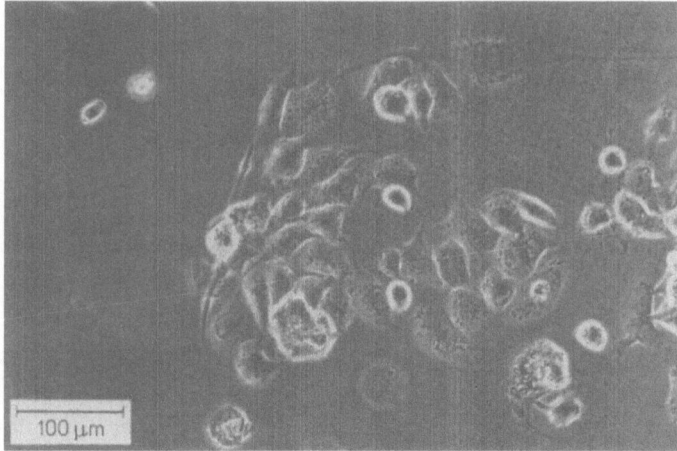

Fig. 7. Neuroblastoma cells derived from the mouse cell line C 1300

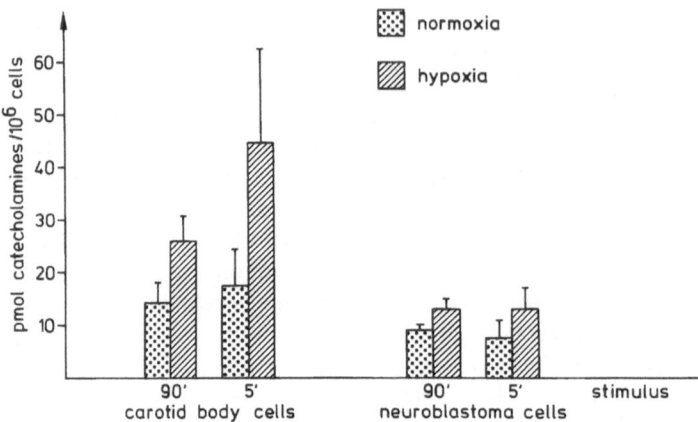

Fig. 8. Effect of hypoxia on catecholamine synthesis. Cells were incubated in medium (1 ml minimum essential medium + 16 ml HEPES-buffered saline, pH 6.0, concentration of labeled + unlabeled tyrosine was 40 nmol/l) containing ^{14}C tyrosine (3 μCi in 200 μl; New England Nuclear, Boston, Mass.) and 10^{-4} mol/l 6-methyltetrahydropterine (MPH$_4$; Sigma, St. Louis, Mo.). At the end of the experiment, cells were digested in 50 μl 0.2 N CH$_3$ COOH and catecholamines were separated by thin-layer chromatography, stained with potassium ferricyanide, and the radioactivity of the spots was determined in a scintillation counter. Means of six experiments ± SEM are shown

Table 1. Effect of hypoxia on catecholamine synthesis (pmol/10^6 cells). The length of the stimulation was 5 or 90 min. Incubation in ^{14}C tyrosine lasted for 90 min in all experiments. Means of six experiments ± SEM are shown

	Carotid body cells			Neuroblastoma cells		
	Normoxia	Stimulation time (min)	Hypoxia	Normoxia	Stimulation time (min)	Hypoxia
Catecholamine	1.9 ± 1.0	5	2.8 ± 1.2	0.5 ± 0.1	5	0.8 ± 0.3
synthesis[a]	0.8 ± 0.1	90	1.4 ± 0.3	0.6 ± 0.04	90	0.8 ± 0.05
Tyrosine	924 ± 137	5	1293 ± 220	1100 ± 385	5	1601 ± 420
	1240 ± 211	90	1550 ± 215	1227 ± 223	90	1369 ± 168
DOPA	11.8 ± 7.2	5	24.4 ± 17.8	0.8 ± 0.4	5	6.9 ± 3.9
	5.1 ± 1.4	90	9.6 ± 2.3	3.0 ± 0.7	90	4.1 ± 1.2
DA	4.3 ± 2.1	5	18.4 ± 12.0	1.2 ± 0.7	5	0.2 ± 0.2
	6.3 ± 2.2	90	12.7 ± 4.9	0.1 ± 0.1	90	0.7 ± 0.3
DOPAC	1.5 ± 0.5	5	1.9 ± 0.83	5.6 ± 3.2	50	5.9 ± 2.3
	2.9 ± 0.4	90	3.9 ± 0.3	6.0 ± 0.7	90	8.3 ± 1.1

DOPA, dihydroxyphenylalanine; DA, dopamine; DOPAC, dihydroxyphenylacetic acid
[a] In % of total label tyrosine

significant rise in catecholamine synthesis after hypoxic stimulation. After a longer hypoxic stimulation of 90 min, the concentration of radioactive catecholamines in carotid body cells was reduced compared to the 5-min stimulus. With 26.1 pmol ± 5.1/10^6 cells after 90 min of hypoxia, only 58% of labeled catecholamines determined after the short stimulus were found. This reduction was due to lower DOPA and dopamine content. As rabbit glomera secrete dopamine under hypoxia (Fidone 1982), it is possible that only a small amount of the new synthesized dopamine remained in the cells after a stimulation time of 90 min.

Taken together the results described above suggest that hypoxia causes a rise in intracellular calcium in carotid body cells as well as an increase in catecholamine synthesis. These conclusions are supported by the findings of Nose et al. (1985), who found in a pheochromocytoma cell line (PC 12) that Ca^{2+} influx elicited by various maneuvers stimulates phosphorylation of tyrosine hydroxylase, which is accompanied by an activation of the enzyme. As tyrosine hydroxylase is the rate-limiting step in catecholamine synthesis, this is the probable site of regulation. From studies on rabbit carotid bodies in vitro (Fidone et al. 1982) and cultured cells from adult rat carotid bodies (Fishman et al. 1985), it is known that this organ secretes dopamine under hypoxia. That extracellular calcium plays a role in this process is also shown by Delpiano (in this volume), who found an inhibiton of the chemosensory nerve response to hypoxia in the absence of calcium in the medium. Hence, calcium influx stimulated by hypoxia triggers both catecholamine secretion and synthesis in carotid body cells.

In summary, we can say that cells of the carotid body in primary culture show morphological and biochemical characteristics similar to those in vivo. Furthermore, they are chemosensitive to oxygen reduction. The exchangeable Ca^{2+} pool increases under hypoxia by 170%. A similar value was found in preliminary experiments in which acetylcholine was used to stimulate the type-I cells (Pietruschka 1985). As glioma cells do not show a comparable reaction to hypoxia, the change in cellular calcium metabolism seems to be a specific process coupled to the chemoreception. Moreover, catecholamine synthesis is stimulated by hypoxia. This effect is more distinct after a short stimulus of 5 min, which induces an increase of labeled catecholamines of 255%, whereas neuroblastoma cells only show a slight and nonsignificant stimulation.

References

Barnes EM, Mandel P (1981) Calcium transport by primary cultured neuronal and glial cells from chick embryo brain. J Neurochem 36: 82–85

Falck B, Hillarp NA, Thieme G, Torp A (1962) Fluorescence of catecholamines and related compounds condensed with formaldehyde. J Histochem Cytochem 10: 348–354

Fidone S, Gonzalez C, Yoshizaki K (1982) Effects of low oxygen on the release of dopamine from the rabbit carotid body in vitro. J Physiol (Lond) 333: 93–110

Fishman MC, Greene WL, Platika D (1985) Oxygen chemoreception by carotid body cells in culture. Proc Natl Acad Sci USA 82: 1448–1450

Fleming RM, Clark WG (1970) Quantitative thin-layer chromatographic estimation of labeled dopamine and norepinephrine, their precursors and metabolites. J Chromatogr 52: 305–312

Haycock JW, Meligeni JA, Bennett WF, Waymire JC (1982) Phosphorylation and activation of tyrosine hydroxylase mediate the acetylcholine-induced increase in catecholamine biosynthesis in adrenal chromaffin cells. J Biol Chem 257: 12641–12648

Nose PS, Griffith LC, Schulman H (1985) Ca^{2+}-dependent phosphorylation of tyrosine hydroxylase in PC12 cells. J Cell Biol 101: 1182–1190

Pietruschka F (1974) Cytochemical demonstration of catecholamines in cells of the carotid body in primary tissue culture. Cell Tiss Res 151: 317–321

Pietruschka F (1985) Calcium influx in cultured carotid body cells is stimulated by acetylcholine and hypoxia. Brain Res 347: 140-143

Pietruschka F, Schäfer D (1976) Fine structure of chemosensitive cells (glomus caroticum) in tissue culture. Cell Tissue Res 168: 55-63

Pietruschka F, Schäfer D, Lübbers DW (1977) Reaction of cultured carotid body cells to different concentrations of oxygen and carbon dioxide. In: Acker H et al. (eds) Chemoreception in the carotid body. Springer, Berlin Heidelberg New York, pp 86-91

Rasmussen H, Barrett PQ (1984) Calcium messenger system: an integrated view. Physiol Rev 64: 939-984

III. Heart and Circulation

PO$_2$-Induced Changes of Membrane Potential and Tension in Vascular Smooth Musculature

G. Siegel[1] and J. Grote[2]

[1] Institute of Physiology, Biophysical Research Group, Free University of Berlin, Arnimallee 22, 1000 Berlin 33, Germany
[2] Institute of Physiology, University of Bonn, Nußallee 11, 5300 Bonn 1, FRG

Introduction

That a decrease in oxygen tension leads to vasodilatation has been known from the literature for a long time. Mechanical measurements demonstrate that hypoxia effects relaxation of vascular strips [8, 9, 10, 33], vasodilatation in perfused vessels [5, 6, 7, 11], or an increase in blood flow [3, 4, 13, 17, 19, 32]. Electrophysiological investigations during oxygen deficiency, especially intracellular measurements of the membrane potential of vascular smooth muscle cells, do not exist. It has been reported merely that a PO$_2$ decrease in the rat portal vein restricts spontaneous spike discharges and finally causes their complete cessation [16]. Johansson and Somlyo [18] supposed that vasodilatation with hypoxia possibly has an electrophysiological correlate.

The regulation of vascular tone is a question of the $[Ca^{2+}]_i$ activity. Presuming electromechanical coupling in a graded depolarization type of smooth muscle, relaxation should be linked to hyperpolarization of the cell membrane [26]. This hyperpolarization should lead to a fall in the free, intracellular Ca^{2+} concentration. Mechanisms which have been discussed up to now for a fall of $[Ca^{2+}]_i$, have so far been of a speculative nature [27]:

1. Under steady-state conditions, Ca^{2+} influx and efflux are equal. Hyperpolarization leads to a decrease in Ca^{2+} permeability. This means a predominance of the Ca^{2+} efflux.
2. Hyperpolarization leads to an increase of the Na^+ driving force (V – E$_{Na}$). This results in an augmentation of the Ca^{2+} efflux on grounds of the Na^+-Ca^{2+} exchange [2].
3. Hyperpolarization leads to adsorption of Ca^{2+} ions to intracellular or membranaceous, specific, or unspecific binding sites. Thus, it is not astonishing that investigations of changes in membrane potential under oxygen deficiency are not yet forthcoming.

Membrane Potential and Vascular Tone with PO$_2$ Variation

The experiments described in the following should elucidate the in vivo role of prostacyclin and endothelium derived relaxant factor (EDRF). Several authors have suggested that hypoxic vasodilatation could be a result of prostacyclin release by endothelial and smooth muscle cells [5, 6, 12, 22, 23, 24, 29, 34, 35]. When measuring membrane potential and tone in function of O$_2$ partial pressure, the following results were found (Fig. 1): In a Krebs solution, which has been equilibrated with carbogen (95% O$_2$ - 5% CO$_2$; PO$_2$ $\approx$ 550 mm Hg), a vessel preparation is predepolarized and precontracted. The systematic decline in oxygen tension of the Krebs solution results in a hyperpolarization of the smooth muscle cell membranes from -65 mV to a maximum of -74 mV at 35 mm Hg PO$_2$ in a dose-

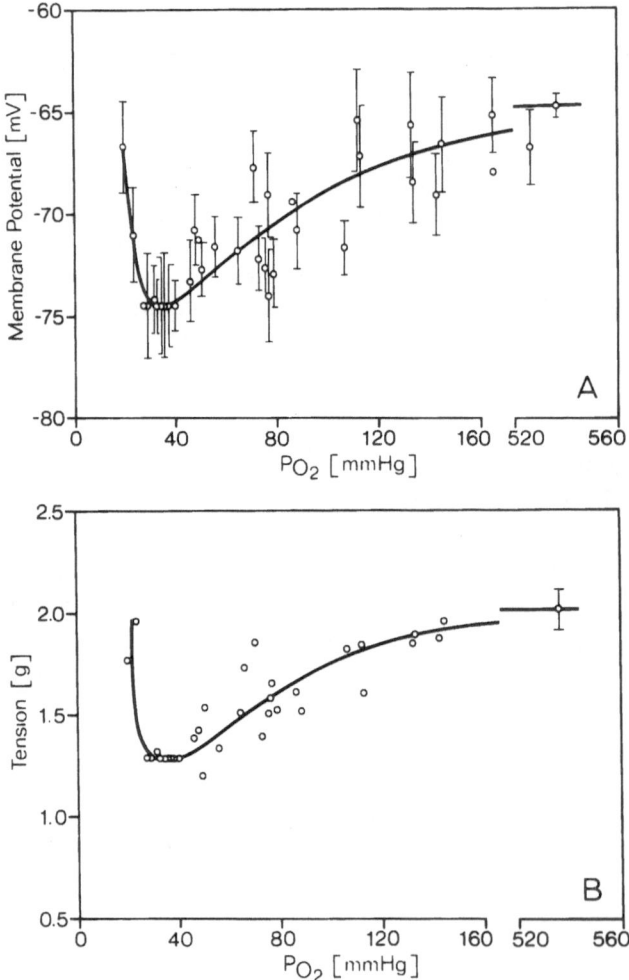

Fig. 1 A, B. Membrane potential (**A**) and mechanical force development (**B**) of isolated carotid segments in response to changes in the oxygen tension of the Krebs solution

dependent manner [14, 15, 28, 29]. Simultaneously, mechanical force development decreases from 2 g to 1.3 g. The half-maximal effect was found at a PO_2 of 87 mm Hg in the sigmoid dose-response curves. When the oxygen tension is lowered further to values of 20 mm Hg, the hyperpolarization recorded in the range of 35 mm Hg $< PO_2 < 535$ mm Hg is reduced, or even depolarization occurs. Mechanically, an increase in tone up to the initial value, or even additional contraction, can be observed.

Figure 1B shows the effect of a change in oxygen tension on mechanical force development. In the range between 535 and 35 mm Hg, the decrease in oxygen partial pressure causes a dose-dependent relaxation of the muscle cells. That is why membrane hyperpolarization and inhibition of muscle tone take a parallel course in function of PO_2. Also below PO_2 values of 35 mm Hg, membrane potential and force development are strictly coupled: The vascular muscle contracts strongly with falling PO_2 and at about 20 mm Hg it reaches its initial tone again, as measured in carbogen Krebs solution. This finding of an effect reversal of force development at very low oxygen partial pressures has already been described and discussed in detail by Rubanyi and Vanhoutte [24]. Quantitatively, it appears that in one experiment the maximal hyperpolarization of the cells amounted to 14 mV, the maximal relaxation to 1.4 g. The average relaxation after 15 min was 0.7 g. A first reaction of the vessel to each stepwise PO_2 decrease occurred after 1-2 min, the total effect was reached after about 10 min.

Since hyperpolarization and relaxation of vascular smooth muscle cells were found also under prostacyclin, their extent reminding of the just described experiments, the investigations with PO_2 variation were repeated under indomethacin. First, however, Fig. 2 demonstrates the effect of iloprost, a stable carbacyclin analogue, on the membrane potential. When iloprost is applied in concentrations between 10^{-9} and 10^{-6} M, the membrane is hyperpolarized in a dose-dependent

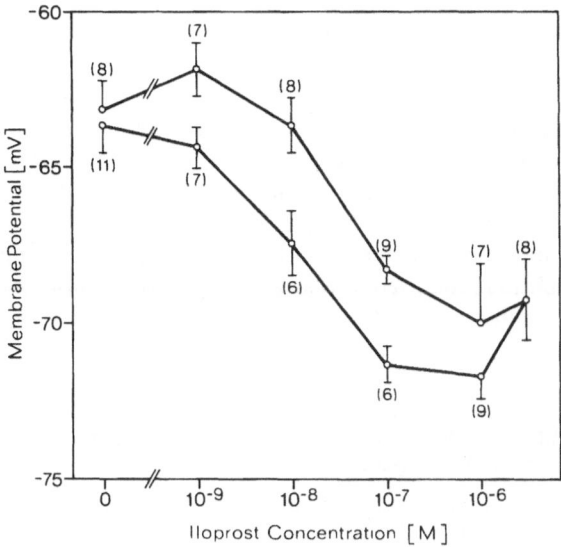

Fig. 2. Membrane potential of vascular smooth muscle cells in dependence on the iloprost concentration in the Krebs solution. Iloprost was applied consecutively for 30 min each at first in rising (more negative potentials), then in falling concentrations (more positive potentials)

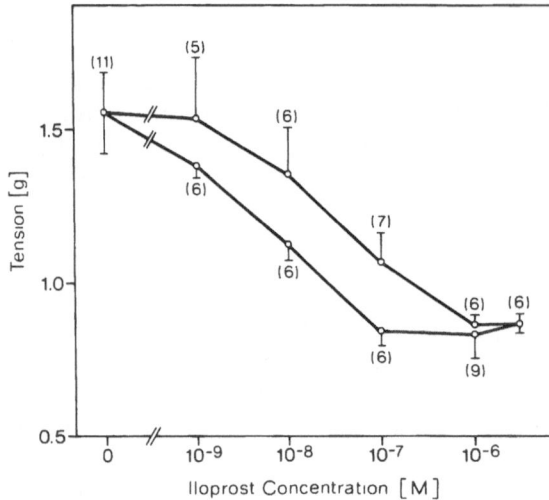

Fig. 3. Dependence of the mechanical tension development of isolated carotid segments on the iloprost concentration in the Krebs solution. Iloprost was applied consecutively for 30 min each at first in rising (lower force values), then in falling concentrations (higher force values)

manner from −63 mV to a maximum of −71 mV [20, 30, 31]. Iloprost concentrations were applied first with increasing, then with decreasing doses for 0.5 h each. When the reversibility was tested, hysteresis occurred. The membrane potentials were more depolarized, the preparations more contracted when the iloprost concentration was diminished again. In principle, all effects were reversible.

According to the theory of electromechanical coupling, a hyperpolarization should lead to vasorelaxation [26]. Figure 3 shows the dose-dependent relaxation of the vascular strip for iloprost concentrations between 10^{-9} and $3 \cdot 10^{-6}$ M. The vessels showed a maximal relaxation under iloprost of 0.7 g. Half-maximal effect on membrane potential and tone was observed with $2 \cdot 10^{-8}$ M concentration.

Blockade of Hypoxia-Induced Hyperpolarization and Vasorelaxation

With the assumption that changes in potential and tension under hypoxia could be mediated by prostacyclin, the effect of indomethacin was tested. Apart from a slight membrane depolarization by 2.1 mV (Fig. 4) and a subsequent augmentation in basal tension by 0.3 g (Fig. 5) [24], indomethacin (10^{-5} M) led to a 20% reduction of the hypoxia-dependent potential and tension alterations. Moreover, the oxygen partial pressure with which maximal hyperpolarization and relaxation occur, is shifted to the significantly higher value of 64 mm Hg PO_2. Although it can

Fig. 4. Membrane potential of the vascular smooth muscle cells in isolated carotid segments treat- ▶ ed by indomethacin (10^{-5} M) as a function of the oxygen tension of the Krebs solution

Fig. 5. Tension developed in isolated carotid segments treated by indomethacin (10^{-5} M) as a ▶ function of the oxygen tension of the Krebs solution

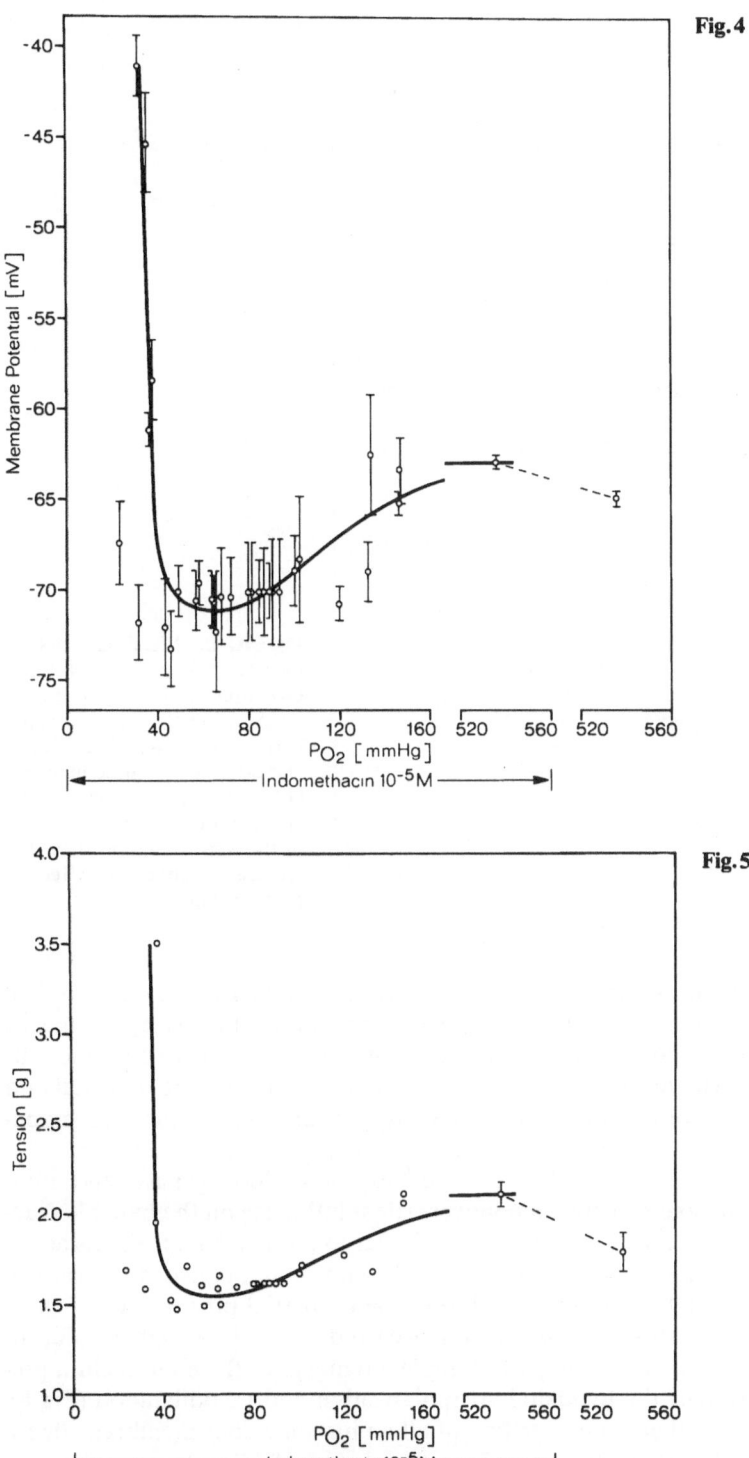

Fig. 4

Fig. 5

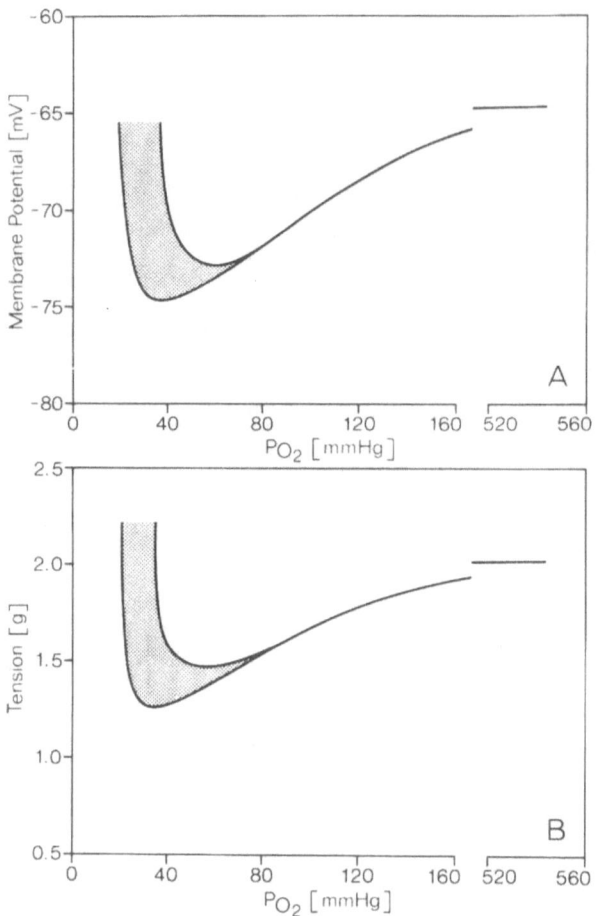

Fig. 6 A, B. Membrane potential (**A**) and mechanical force development (**B**) of isolated carotid segments treated *(upper curves)* or untreated *(lower curves)* by indomethacin $(10^{-5} M)$ in response to changes in the oxygen tension of the Krebs solution. The *dotted area* illustrates the effect of prostacyclin

be supposed that the indomethacin concentration applied leads to a complete blockade of prostaglandin synthesis, the hypoxic vasodilatation is by no means entirely absent under indomethacin. The effects of indomethacin are once more illustrated in Fig. 6. The influence of a reduction in PO_2 on membrane potential and tension is compared for indomethacin-treated (upper curves) and untreated preparations (lower curves). The dotted area represents the effect of prostacyclin.

Even application of BW 755 C $(3.8 \cdot 10^{-5} M)$, a cyclooxygenase and lipooxygenase inhibitor, exerts merely a relatively slight influence on the hypoxic reactivity of the blood vessel (Fig. 7). BW 755 C, similar to indomethacin, effects an insignificant membrane depolarization of 3.9 mV with a simultaneous increase in tone of 0.4 g. Essentially, however, the shift of the O_2 partial pressure, with which maximal hyperpolarization and relaxation occurred, towards a higher value of 57 mm Hg PO_2 is remarkable. Only the complete removal of the endothelium prevents for the most part the hypoxic hyperpolarization and vasodilatation (Fig. 8). Scraping off the endothelium causes first of all an instantaneous membrane depolarization of 4.1 mV and contraction of the vessel strip of 1.3 g. Decrease in oxy-

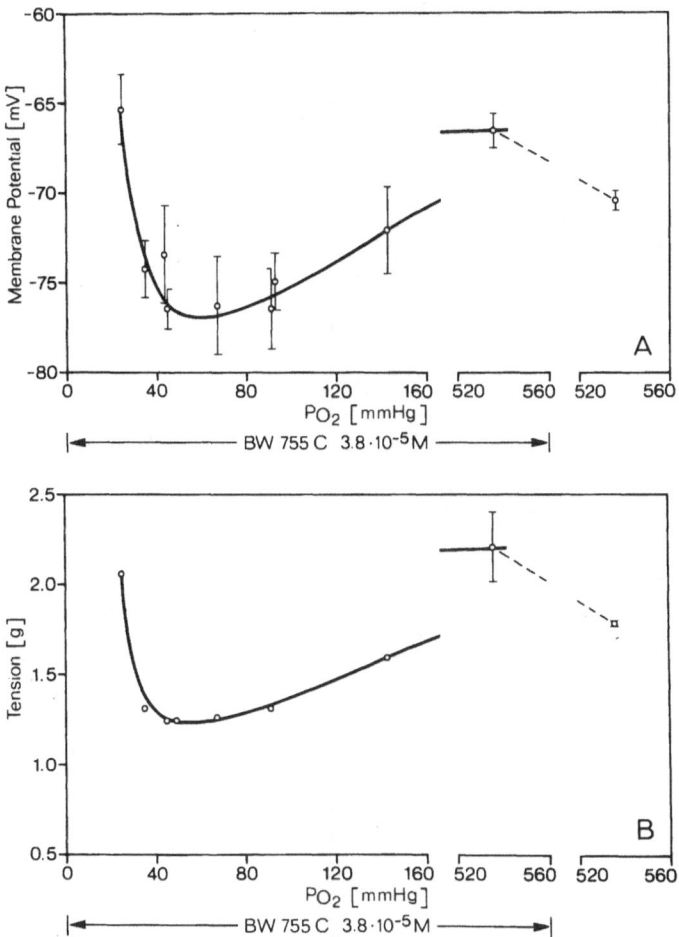

Fig. 7 A, B. Membrane potential (**A**) and tension (**B**) developed in isolated carotid segments treated by BW 755 C ($3.8 \cdot 10^{-5}$ M) as a function of the oxygen tension of the Krebs solution

gen tension down to a value of 69 mm Hg, however, leads only to a slight hyperpolarization of 1.7 mV with simultaneous relaxation of 0.2 g. Below a PO_2 of 69 mm Hg again depolarization and vasoconstriction occur. Therefore, one can assume that about 80% of the effects in potential and tone under a decrease of oxygen tension are to be attributed to the endothelial dilator EDRF.

The present results permit no explanation of the causal effects of hypoxia. According to our results and to statements in the literature, the release of prostacyclin [5, 6, 22, 30], EDRF [5, 6, 12, 23, 24, 34, 35], and oxygen-free radicals [1, 36] is charged causally with the membrane hyperpolarization and vasorelaxation during oxygen deficiency. In case of a direct influence of these substances on the cell membrane, a change of passive membrane properties (increase in K^+ permeability, decrease in Na^+ permeability) or of active transport processes (stimulation of the electrogenic Na^+ outward pump) can be taken into consideration as reason

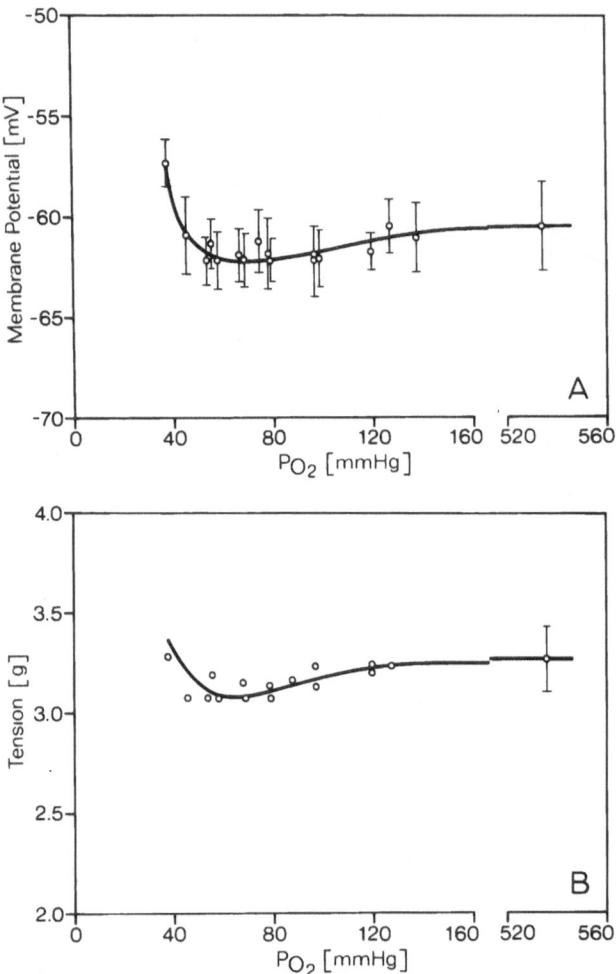

Fig. 8 A, B. Membrane potential (A) and mechanical force development (B) of isolated carotid segments after removal of the endothelium in response to changes in the oxygen tension of the Krebs solution

for the membrane hyperpolarization. For prostacyclin the effector mechanism is cleared up. A strong increase in K^+ permeability and a stimulation of the electrogenic Na^+ outward transport are responsible for the observed hyperpolarization [20, 30, 31].

Electromechanical Coupling

Examining the electromechanical coupling by elimination of the parameter PO_2 from the potential-tone relations results in the lower, sigmoid part of the activation curve for the experimental series without and with indomethacin (Fig. 9). This curve is comparable to the relation under iloprost. In this figure, the inhibition-

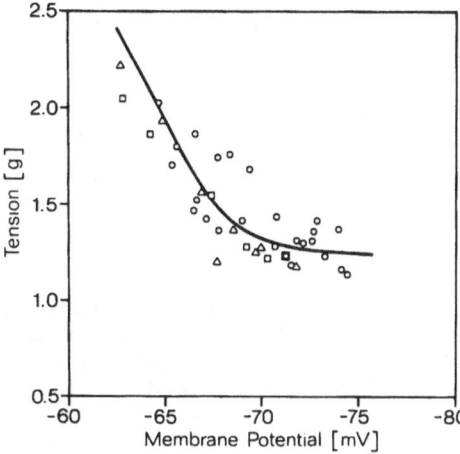

Fig. 9. Dependency of mechanical tension on the membrane potential in isolated carotid segments (stationary activation curve). The membrane potential of normal (O) and preparations incubated with indomethacin (10^{-5} M) (□) was changed by varying the oxygen tension in the Krebs solution. Furthermore, the graph contains measurement values in which the membrane potential was altered by variations of the iloprost concentration (△) in the blood substitute solution

relaxation coupling can be seen, where vasodilatation is correlated with hyperpolarization [20]. The graph represents the hyperpolarized section of the sigmoid, stationary activation curve [28, 29, 30].

To arrive at the whole activation curve, we eliminated the respective effector parameter from diagrams of membrane potential as well as tension development in function of $[H^+]_o$, $[K^+]_o$, $[Ca^{2+}]_o$, noradrenaline, or prostacyclin concentration of the blood substitute solution and plotted the developed tension as function of the membrane potential (Fig. 10). From the sigmoid, stationary activation curve and the measured values, it follows immediately that the same change in tone oc-

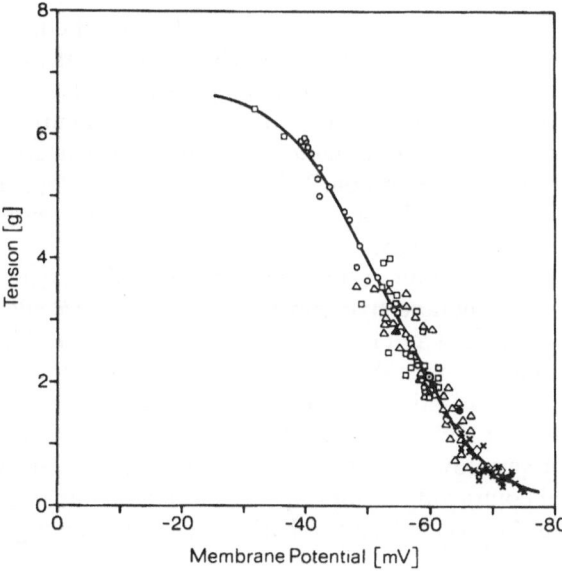

Fig. 10. Dependency of mechanical tension on the membrane potential in isolated carotid arteries. The membrane potential was changed by varying the extracellular concentration of H^+ (△), K^+ (□), Ca^{2+} (●), norepinephrine (O), or prostacyclin (◇), or by lowering the oxygen tension (×). The marked point on the curve (⊙) indicates membrane potential and mechanical force under control conditions

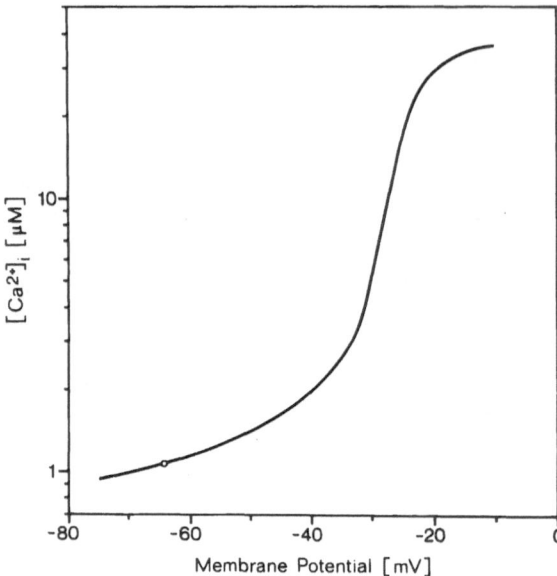

Fig. 11. Ionized, intracellular Ca^{2+} concentration as a function of the membrane potential of arterial smooth muscle cells. The marked point on the curve (O) represents $[Ca^{2+}]_i$ under physiological conditions, the membrane resting potential being -64.4 mV. (The curve was obtained by using data from Fig. 8.14 in Rüegg [25])

curs with a defined change in membrane potential, no matter from which effector influence the latter originates [26, 28, 29]. Thus, this curve serves as strong evidence for the existence of electromechanical coupling in vascular smooth muscle, at least in the carotid artery. The marked point on the curve represents the membrane potential and tension under normal conditions. In the linear range of the characteristic, a potential shift of 5 mV corresponds to a change in tone of 1 g. This linear part comprises about 20 mV and thus 70% of the force developed.

In conclusion, we would like to enter into discussion of the question of how membrane hyperpolarization leads to vascular relaxation. Recently, with the help of patch-clamp techniques, a second category of Ca^{2+} channels was found, the activation curve of which lies between -80 mV and -40 mV [21]. These Ca^{2+} channels are closed voltage-dependently with hyperpolarization. Ca^{2+} inward current eliciting contraction or inward current of trigger calcium are suppressed. The consequence is a fall in intracellular Ca^{2+} concentration. This is represented in Fig. 11 in dependence on the membrane potential. Depolarization leads to an increase, hyperpolarization to a decrease of Ca^{2+} activity.

It can be concluded that the vasodilatation occurring in arteries under oxygen deficiency can be attributed to a membrane hyperpolarization of the vascular smooth muscle cells. For hyperpolarization and relaxation, 20% of the responsibility lies with prostacyclin release and 80% with the release of the endothelial vasodilator EDRF. Obviously, the hyperpolarization is brought about by an increase of passive K^+ permeability. Whether the membrane depolarization and contraction of vessel segments observed with very low oxygen partial pressures is caused by release of an endothelium-derived contracting factor [24] or by noradrenaline from sympathetic nerve endings cannot be decided on the basis of existing investigations.

Acknowledgements. The authors thank Mrs. Ch. Fuhrmann for her excellent technical assistance and Mr. H. Ewald from the mechanical workshop for his constant help. We appreciate the outstanding graphic representations by Mrs. M. Krawczynski as well as the translation and typing of the manuscript by Mrs. E. Gaebel.

References

1 Arfors K-E (1988) Free radicals. In: Tsuchiya M (ed) Proc 4th World Congr Microcirc. Elsevier, Amsterdam (in press)
2 Blaustein MP (1976) Sodium-calcium exchange and the regulation of cell calcium in muscle fibers. Physiologist 19: 525–540
3 Block AJ, Feinberg H, Herbaczynska-Cedro K, Vane JR (1975) Anoxia-induced release of prostaglandins in rabbit isolated hearts. Circ Res 36: 34–42
4 Busija DW (1984) Sympathetic nerves reduce cerebral blood flow during hypoxia in awake rabbits. Am J Physiol 247: H446–H451
5 Busse R, Förstermann U, Matsuda H, Pohl U (1984) The role of prostaglandins in the endothelium-mediated vasodilatory response to hypoxia. Pflugers Arch 401: 77–83
6 Busse R, Pohl U, Kenner C, Klemm U (1983) Endothelial cells are involved in the vasodilatory response to hypoxia. Pflugers Arch 397: 78–80
7 Carrier JRO, Walker JR, Guyton AC (1964) Role of oxygen in autoregulation of blood flow in isolated vessels. Am J Physiol 206: 951–954
8 Chang AE, Detar R (1980) Oxygen and vascular smooth muscle contraction revisited. Am J Physiol 238: H716–H728
9 Coburn RF, Grubb B, Aronson RD (1979) Effect of cyanide on oxygen tension-dependent mechanical tension in rabbit aorta. Circ Res 44: 368–378
10 Detar R (1980) Mechanism of physiological hypoxia-induced depression of vascular smooth muscle contraction. Am J Physiol 238: H761–H769
11 Duling BR (1974) Oxygen sensitivity of vascular smooth muscle. II. In vivo studies. Am J Physiol 227: 42–49
12 Furchgott RF, Zawadzki JV (1980) The obligatory role of endothelial cells in the relaxation of arterial smooth muscle by acetylcholine. Nature 288: 373–376
13 Grote J, Schubert R (1982) Regulation of cerebral perfusion and PO_2 in normal and edematous brain tissue. In: Loeppky JA, Riedesel ML (eds) Oxygen transport to human tissues. Elsevier/North-Holland, Amsterdam, pp 169–178
14 Grote J, Siegel G, Adler A, Zimmer K, Müller R (1987) The effect of hypoxia on the electromechanical properties of the canine carotid artery. In: Cervós-Navarro J, Ferszt R (eds) Stroke and microcirculation. Raven, New York, pp 51–56
15 Grote J, Siegel G, Zimmer K, Adler A (1987) Membranpotential und Tonus der Gefäßmuskulatur in Abhängigkeit vom O_2-Partialdruck. Z Kardiol 76: 35
16 Hellstrand P, Johansson B, Norberg K (1977) Mechanical, electrical, and biochemical effects of hypoxia and substrate removal on spontaneously active vascular smooth muscle. Acta Physiol Scand 100: 69–83
17 Jackson WF, Duling BR (1983) The oxygen sensitivity of hamster cheek pouch arterioles. In vitro and in situ studies. Circ Res 53: 515–525
18 Johansson B, Somlyo AP (1980) Electrophysiology and excitation-contraction coupling. In: Bohr DF, Somlyo AP, Sparks JRHV (eds) Handbook of Physiology, Sect 2, The cardiovascular system, Vol II, Vascular smooth muscle. American Physiological Society, Bethesda, pp 301–323
19 Kontos HA, Wei EP, Raper AJ, Rosenblum WI, Navari RM, Patterson JL (1978) Role of tissue hypoxia in local regulation of cerebral microcirculation. Am J Physiol 234: H582–H591
20 Litza B, Siegel G (1987) Inhibition-relaxation coupling in vascular smooth muscle. Fed Proc 46: 507
21 Loirand G, Pacaud P, Mironneau C, Mironneau J (1986) Evidence for two distinct calcium channels in rat vascular smooth muscle cells in short-term primary culture. Pflugers Arch 407: 566–568

22 McCalden TA, Nath RG, Thiele K (1983) Prostacyclin and vasodilator mechanisms in the cerebral circulation. Blood Vessels 20: 202
23 Palmer RMJ, Ferrige AG, Moncada S (1987) Nitric oxide release accounts for the biological activity of endothelium-derived relaxing factor. Nature 327: 524-526
24 Rubanyi GM, Vanhoutte PM (1985) Hypoxia releases a vasoconstrictor substance from the canine vascular endothelium. J Physiol (Lond) 364: 45-56
25 Rüegg JC (1986) Calcium in muscle activation. Springer, Berlin Heidelberg New York
26 Siegel G (1986) Membranphysiologische Grundlagen der peripheren Gefäßregulation. Physiol aktuell 1: 31-52
27 Siegel G, Ehehalt R, Koepchen HP (1978) Membrane potential and relaxation in vascular smooth muscle. In: Vanhoutte PM, Leusen I (eds) Mechanisms of vasodilatation. Karger, Basel, pp 56-72
28 Siegel G, Grote J (1987) Hypoxia effects hyperpolarization and relaxation in canine vascular smooth muscle. Fed Proc 46: 507
29 Siegel G, Grote J, Zimmer K, Adler A, Litza B (1988) Electro-physiological effects of hypoxia on vascular smooth muscle. In: Vanhoutte PM (ed) Vasodilatation. Raven, New York (in press)
30 Siegel G, Stock G, Schnalke F, Litza B (1987) Electrical and mechanical effects of prostacyclin in the canine carotid artery. In: Gryglewski RJ, Stock G (eds) Prostacyclin and its stable analogue iloprost. Springer, Berlin Heidelberg New York, pp 143-149
31 Siegel G, Thiel M, Schnalke F, Litza B, Adler A, Stock G (1986) Prostacyclin und Vasodilatation. Klin Wochenschr 64: 1156-1157
32 Sylvester JT, Scharf SM, Gilbert RD, Fitzgerald RS, Traystman RJ (1979) Hypoxic and CO hypoxia in dogs: hemodynamics, carotid reflexes, and catecholamines. Am J Physiol 236: H22-H28
33 Vanhoutte PM (1976) Effects of anoxia and glucose depletion on isolated veins of the dog. Am J Physiol 230: 1261-1268
34 Vanhoutte PM (1987) The end of the quest? Nature 327: 459-460
35 Vanhoutte PM, Rubanyi GM, Miller VM, Houston DS (1986) Modulation of vascular smooth muscle contraction by the endothelium. Annu Rev Physiol 48: 307-320
36 Wolin MS, Rodenburg JM, Messina EJ, Kaley G (1987) Oxygen metabolites and vasodilator mechanisms in rat cremasteric arterioles. Am J Physiol 252: H1159-H1163

Possible Function of Endothelial Cells as Oxygen Sensors*

U. Pohl, R. Busse, J. Galle, and E. Bassenge

Institut für Angewandte Physiologie der Universität, Hermann-Herder-Straße 7, 7800 Freiburg, FRG

Introduction

The matching of oxygen supply with oxygen demand is one of the main features of circulatory control. Alterations of oxygen supply due to changes in oxygen tension of the arterial blood rapidly induce changes in cardiac function and peripheral resistance to counterbalance changes of tissue oxygenation [24]. Experiments suggest, that in addition to chemoreception by carotid bodies such changes of pO_2 are sensed in the peripheral vasculature as well. Small arteries in situ showed vasomotor responses to changes in local oxygen tension even if they were freed from adjacent tissue [10]. Further evidence can be derived from the fact that, during rapid changes of blood pO_2, changes of vascular resistance precede those of tissue oxygenation [19]. However, the nature and localization of oxygen-sensing structures in the vascular wall are not fully understood. There is now a growing body of evidence that endothelial cells which line the luminal surface of blood vessels exert a potent vasomotor influence [7], which is mainly due to the release of autacoids such as endothelium-derived relaxing factor (EDRF) [3, 7] and, to a lesser degree, by prostaglandins [2, 17]. Therefore it was tested whether the endothelium plays a role in the vasomotor response of arteries to changes in pO_2 and whether this may be mediated by the release of endothelial autacoids. A pO_2-dependent alteration of the release of endothelium-derived vasoactive autacoids would imply that vascular endothelium potentially acts as an oxygen sensor in the regulation of tissue perfusion and oxygenation.

Methods

To determine the effects of hypoxia (pO_2 range, 20–40 mm Hg) on endothelial cells, a model was developed which allowed the pO_2 at the endothelial side of blood vessels to decrease, while the pO_2 at the adventitial side was maintained at high levels. Isolated segments of canine, rabbit, hamster, and rat arteries were cannulated at both ends and mounted in an organ bath (Tyrode's solution) gassed to a

* Supported by the Deutsche Forschungsgemeinschaft (Grant Bu 436/3-2)

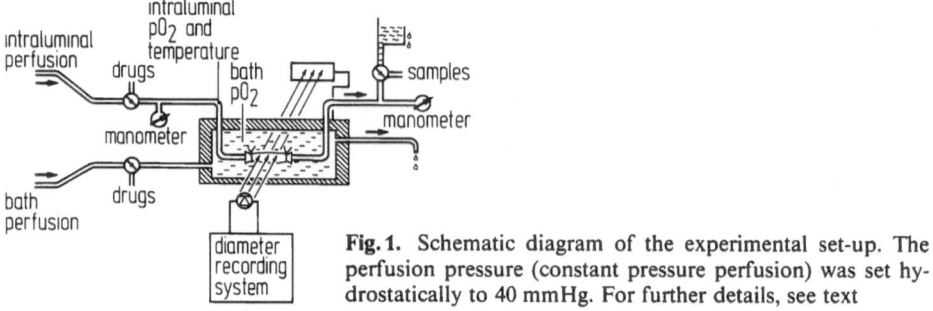

Fig. 1. Schematic diagram of the experimental set-up. The perfusion pressure (constant pressure perfusion) was set hydrostatically to 40 mm Hg. For further details, see text

pO_2 of 300 mm Hg. Intraluminally, these segments were perfused at a rate of 40 ml/h with Tyrode's solution gassed to a pO_2 of either 130 mm Hg (normoxia) or 20–40 mm Hg (hypoxia; Fig. 1). This approach differs from the usual experimental preparation for studying vascular strips in an organ bath where both endothelial *and* adventitial sides are subjected to the same changes of pO_2. The diameter changes of the arterial segments precontracted with norepinephrine (ED30) in response to hypoxia were recorded by a photoelectric device. The role of endothelial cells was investigated by comparison of segments with intact endothelium and segments with the endothelium removed mechanically. The potential role of EDRF in vasodilations was studied by means of the EDRF inhibitors hemoglobin (5 μM) and dithiothreitol (0.2 mM). In separate experiments on vascular segments and cultured endothelial cells, the effects of changes in endothelial pO_2 on the release of prostacyclin were studied measuring the amounts of its stable hydrolysate, 6-keto $PGF_{1\alpha}$ by radioimmunoassay. The cultured cells were grown on microcarrier beads and, for the investigation of pO_2 effects, packed in small columns perfused with Tyrode's solution. The release of EDRF from these cells was determined by means of an endothelium-denuded assay segment, which was superfused with the effluent from the cell column. Culture conditions for the endothelial cells and details of the methods applied have been described earlier [2, 3, 15].

Results

In segments of canine femoral artery, rat tail artery, hamster carotid artery and rabbit aorta, luminal hypoxia (20–40 mm Hg) induced a vasodilation in 60%–80% of the segments (Table 1; Fig. 2). In any given vessel significantly higher dilations were observed at pO_2 values of 20 mm Hg than at values of 40 mm Hg (mean increase in rabbit aortae, 32% ± 12%; $P < 0.05$). By contrast, in segments without endothelium, no changes in vascular diameter were observed during selective luminal hypoxia.

In rat tail arteries, the hypoxia-induced dilation was associated with an enhanced release of prostacyclin from the vessel segments (Fig. 2). Inhibiton of cyclooxygenase by indomethacin (20 μM) in these segments resulted in abolition of both release of prostacyclin and hypoxic dilations. In segments of rat tail artery without endothelium neither a basal nor a hypoxia-induced release of prostacyclin

Table 1. Vasomotor responses of isolated arteries to endothelial hypoxia (pO_2, 20–40 mm Hg)

	Tail artery Rat	Femoral artery Dog	Carotid artery Hamster	Thoracic aorta Rabbit
Outer diameter (OD) (μm) (nonstimulated)	849 ± 17	1780 ± 35	765 ± 19	3864 ± 152
Precontraction (PC) (% of OD) (norepinephrine)	16 ± 1	18 ± 3	18 ± 2	12 ± 1
Hypoxia-induced dilation (% of PC)				
Intact endothelium	39 ± 4 (16/23)	19 ± 3 (14/18)	10 ± 2 (10/13)	12 ± 2 (33/56)
Endothelium-denuded	0.6 ± 0.2 (14)	1.3 ± 0.5 (6)	0.6 ± 0.7 (10)	0.6 ± 0.1 (32)

Segments with endothelium:
(x/y) indicate number of experiments in which hypoxia-induced dilation was observed *(x)*/total number of experiments *(y)*
Segments without endothelium:
(n) = total number of experiments

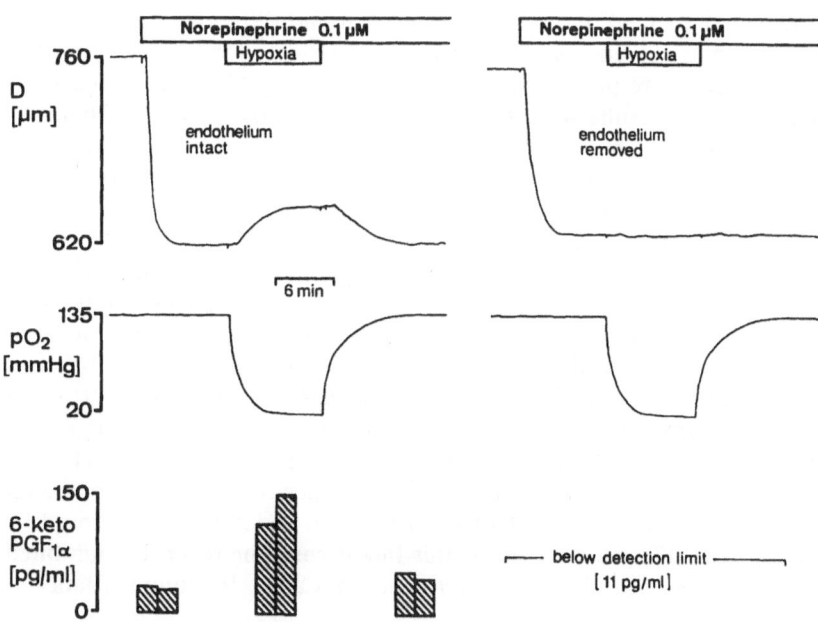

Fig. 2. Dilation of a rat tail artery with intact endothelium *(left upper panel)*, during reduction of the PO_2 of the luminal perfusate *(middle panel)*. Hypoxic perfusion is associated with an enhanced release of PGI_2 from the segment, measured as concentration of its stable hydrolysate 6-keto-$PGF_{1\alpha}$ in samples (2 min) of the effluent *(lower panel)*. In the absence of endothelium, the same decrease of luminal PO_2 neither elicits a dilation nor a release of PGI_2. D = outer diameter of the segment. Initial contraction ("precontraction") is elicited by administration of norepinephrine to the organ bath as indicated

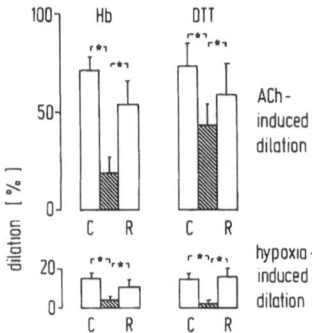

Fig. 3. Effect of the EDRF inhibitors hemoglobin *(Hb, 5 µM)* and dithiothreitol *(DTT, 0.2 mM)* on acetylcholine-induced (ACh, 1 µM) and hypoxia-induced (PO_2 20–40 mmHg) dilations in rabbit aortic segments (endothelium intact). Compared to control responses *(C)* these compounds significantly reduce both dilator responses *(hatched bars)*. This inhibition is reversible as indicated by the recovery *(R)* values after washout of the compounds. Mean ± SEM of 11 (Hb) and 4 (DTT) experiments; *P ≤ 0.05

was detected (Fig. 2). In some canine femoral arteries, however, the suppression by indomethacin of the hypoxic dilation was incomplete, though no release of prostacyclin from these vessels could be detected. This finding indicated that other mechanisms or factors than prostaglandins might be involved in eliciting the endothelium-mediated pO_2-dependent dilation in these vessels.

Therefore, experiments were performed in isolated segments of rabbit aorta, which are known to be practically insensitive to prostacyclin [6]. In these vessels indomethacin did not affect the dilation in response to luminal hypoxia. However, in the presence of the EDRF inhibitors hemoglobin and dithiothreitol it was significantly attenuated or even abolished. In the presence of these inhibitors the EDRF-mediated dilator responses to the acetylcholine (1 µM) were similarly inhibited (Fig. 3). Similar results were obtained in a preliminary study on hamster carotid arteries (data not shown). Intraluminal perfusion of the aortic segments with nominally Ca^{2+}-free Tyrode's solution, with Ca^{2+} still present in the organ bath, reversibly inhibited the hypoxia-induced dilation (from 11.0% ± 2.3% to 2.5% ± 2.5%; $P < 0.01$; $n = 4$).

The cultured endothelial cells showed a behaviour similar to that of native endothelium in vascular segments in response to hypoxia. With reduction of the pO_2 of the perfusate of cells packed into small columns, the prostacyclin release from these cells gradually increased (by 122% ± 88%; $P < 0.05$). This was reversible with normoxic reperfusion. Neither the cellular ATP levels (5.5 ± 0.6 nmol/10^6 cells) nor the cellular energy charge (calculated from ATP, ADP, and AMP levels) showed significant changes after 20 min perfusion with a pO_2 of 25 mm Hg. In separate bioassay experiments, performed in the presence of indomethacin, the release of a nonprostanoid-relaxing factor from hypoxic endothelial cells could be observed ($n = 5$). The dilation induced by this factor could be reversibly inhibited by administration of hemoglobin to the effluent from the cell columns (data not shown).

Discussion

Endothelial cells can decisively affect the vascular tone, which implies an important role for these cells in the regulation of blood flow. They are stimulated to release vasoactive autacoids in response to many endogenous and exogenous substances [for review see 3, 7] as well as to mechanical stimuli associated with pulsatile flow conditions and changes in flow [20, 21]. The present experiments on isolated arteries and cultured endothelium demonstrate that endothelial cells also respond to changes of luminal pO_2 and can play a distinct role in the pO_2-dependent vasomotion of arteries. This hypoxia-induced endothelium-dependent vasodilation appears to be mediated by the release of vasoactive autacoids from the endothelium. In the rat tail artery prostacyclin seems to be the most important mediator [3]. In the other types of arteries cyclooxygenase products were released to a greater or lesser extent (unpublished data), but they played either no or only a limited role as mediators. The inhibition of the hypoxic dilations by hemoglobin and by dithiothreitol, which have both been shown to be potent inhibitors of EDRF in saline perfused tissues [8, 15, 16], suggest that in these vessels EDRF may act as mediator of the endothelium-dependent vascular response to hypoxia. The bioassay experiments demonstrating the release of a nonprostanoid, Hb-sensitive factor from cultured endothelial cells during hypoxic perfusion are also consistent with this. However, since EDRF is more rapidly inactivated at higher pO_2 levels [3, 15] – probably due to its radical nature [18] – it is possible that the hypoxia-induced dilation results not only from an enhanced EDRF release, but in part also from an attenuated EDRF inactivation during hypoxia.

No evidence was found for the release of an endothelial-constricting factor during hypoxia, which has been described by Rubanyi et al. [22]. However, the decrease of pO_2 produced in the studies here was not as severe as the complete anoxia studied by these authors, which abolishes the production of EDRF [4, 9]. The removal of continuously released EDRF might well contribute to a net constrictor response of isolated vessels during anoxia.

The mechanism of the enhanced release of autacoids from endothelial cells at low levels of pO_2 is still unclear. The precursor of prostacyclin, arachidonic acid, accumulates during hypoxia [13]. It has been proposed that this might be due to a pO_2-dependent activation of phospholipase A_2 or an impaired reacylation of arachidonic acid secondary to a hypoxia-induced lack of ATP [11]. Experiments performed on highly metabolically active MDCK cells suggest that a relative lack of ATP near the cell membrane may be responsible for an enhanced PGI_2 release from these cells [12]. We found no reduction of ATP content or of energy charge in cultured endothelial cells with hypoxic perfusion. This does not argue in favor of a role of ATP-depletion, although measurements of mean cellular levels of ATP (and ADP and AMP) may not be sensitive enough to answer this question. EDRF has been recently identified to be the NO-radical [18], but nothing as yet is known about its cellular metabolic pathway. Therefore, an explanation with regard to a direct or indirect role of oxygen in its production would be purely speculative at this stage. However, our studies demonstrated that both autacoids are released in parallel during hypoxic challenge of the cells, and the production and/or release of both autacoids is calcium dependent [3, 9, 15, 23]. Therefore the increase of intracellular free calcium which we observed in hypoxic perfused-cultured endothelial cells [14] might be a common early step in the reaction cascade of enhanced autacoid release during hypoxia.

In summary, our experiments demonstrate that endothelial cells act as oxygen sensors in the vascular wall, which signal changes of pO_2 by alterations of their release of vasoactive autacoids. It is beyond the scope of these in vitro studies to evaluate the significance of this endothelial oxygen-sensor function in relation to other centrally and locally generated signals during alterations of tissue oxygen supply. Moreover, since the experiments were performed in conductance arteries, no definite conclusions can be drawn as to the role of the endothelial cells in the local responses of resistance vessels to pO_2 changes. However, the "hypoxic" pO_2 values as studied in our experiments do naturally occur in these small arteries and arterioles [5]. Therefore, it is tempting to speculate that in these small vessels the autacoid release from vascular endothelium is continuously stimulated by the low pO_2, which may represent an important component in the local balance of vasoconstrictor and dilator influences on resistance vessels.

References

1 Busse R, Pohl U, Kellner C, Klemm U (1983) Endothelial cells are involved in the vasodilatory response to hypoxia. Pflugers Arch 397: 78–80
2 Busse R, Förstermann U, Matsuda H, Pohl U (1984) The role of prostaglandins in the endothelium-mediated vasodilatory response to hypoxia. Pflugers Arch 401: 77–83
3 Busse R, Trogisch G, Bassenge E (1985) The role of endothelium in the control of vascular tone. Basic Res Cardiol 80: 475–490
4 DeMey JG, Vanhoutte PM (1978) Oxygen-dependency of the acetylcholine-induced relaxation in vascular smooth muscle. Arch Int Pharmacodyn Ther 234: 339
5 Duling BR, Berne RM (1970) Longitudinal gradients in periarteriolar oxygen tension. A possible mechanism for the participation of oxygen in local regulation of blood flow. Circ Res 27: 669–678
6 Förstermann U, Hertting G, Neufang B (1984) The importance of endogenous prostaglandins other than prostacyclin for the modulation of contractility of some rabbit blood vessels. Br J Pharmacol 81: 623–630
7 Furchgott RF (1984) The role of endothelium in the responses of vascular smooth muscle to drugs. Annu Rev Pharmacol Toxicol 24: 175–197
8 Griffith TM, Edwards DH, Lewis MJ, Newby AC, Henderson AH (1984) The nature of endothelium-derived vascular relaxant factor. Nature 308: 645–647
9 Griffith TM, Edwards DH, Newby AC, Lewis MJ, Henderson AH (1986) Production of endothelium-derived relaxant factor is dependent on oxidative phosphorylation and extracellular calcium. Cardiovasc Res 20: 7–12
10 Jackson WF, Duling BR (1983) The oxygen sensitivity of hamster cheek pouch arterioles. In vitro and in situ studies. Circ Res 53: 515–525
11 Kurtz A, Jelkmann W, Pfeilschifter J, Bauer Ch (1985) Role of prostaglandins in hypoxia-stimulated erythropoietin production. Am J Physiol 249: C3–C8
12 Kurtz A, Pfeilschifter J, Malmström K, Woodson RD, Bauer Ch (1987) Mechanism of NaCl transport-stimulated prostaglandin formation in MDCK cells. Am J Physiol 252: C307–C314
13 Lands WEM, Sauter J, Stone GW (1987) Oxygen requirement for prostaglandin biosynthesis. Prostaglandins Leukotrienes Med 1: 117–120
14 Lückhoff A, Pohl U, Busse R (1986) Increased free calcium in endothelial cells in response to hypoxia and restitution of normoxia. Pflugers Arch 406 [Suppl]: R46
15 Lückhoff A, Busse R, Winter I, Bassenge E (1987) Characterization of vascular relaxant factor from cultured endothelial cells. Hypertension 9: 295–303

16 Martin W, Villani GM, Jothianantan D, Furchgott RF (1985) Selective blockade of endothelium-dependent and glyceryl trinitrate-induced relaxation by hemoglobin and by methylene
   blue in the rabbit aorta. J Pharmacol Exp Ther 232: 708-716
17 Moncada S, Vane JR (1979) The role of prostacyclin in vascular tissue. Fed Proc 38: 66
18 Palmer RJM, Ferrige AG, Moncada S (1987) Nitric oxide release accounts for the biological
   activity of endothelium-derived relaxing factor. Nature 327: 524-526
19 Pohl U, Busse R, Kessler M (1982) Vascular resistance and tissue pO_2 in skeletal muscle during perfusion with hypoxic blood. In: Kenner T, Busse R, Hinghofer-Szalkay H (eds) Cardiovascular system dynamics: models and measurements. Plenum, New York, pp 521-530
20 Pohl U, Busse R, Kuon E, Bassenge E (1986) Pulsatile perfusion stimulates the release of endothelial autacoids. J Appl Cardiol 1: 215-235
21 Pohl U, Holtz J, Busse R, Bassenge E (1986) Crucial role of endothelium in the vasodilator response to increased flow in vivo. Hypertension 8: 37-44
22 Rubanyi GM, Vanhoutte PM (1985) Hypoxia releases a vasoconstrictor substance from the canine vascular endothelium. J Physiol (Lond) 364: 45-56
23 Singer HA, Peach MJ (1982) Calcium- and endothelial-mediated vascular smooth muscle relaxation in rabbit aorta. Hypertension 4: II19-II25
24 Sylvester JT, Scharf SM, Gilbert RD, Fitzgerald RS, Traystman RJ (1979) Hypoxic and CO
   hypoxia in dogs: hemodynamics, carotid reflexes, and catecholamines. Am J Physiol 236:
   H22-H28

Influence of Oxidative Stress on Metabolic and Contractile Functions of Arterial Smooth Muscle

H. Heinle

Physiological Institute (I), Gmelinstr. 5, 7400 Tübingen, FRG

Introduction

Since the highest oxygen pressure within the body is found in oxygenated arterial blood, the vascular tissue of the arteries is under permanent oxidative attack. Besides the important function of oxygen in the energy-providing metabolism of mitochondria, there are several other pathways in the cells which are influenced either directly by oxygen or by its derivatives, i.e., superoxide anion, hydroxyl radical, H_2O_2, etc. (for review see Halliwell 1978; Chance et al. 1979; Del Maestro 1980; Fridovich 1983). These reactive compounds must be considered normal metabolites in aerobic life.

In arterial smooth muscle, the vasoactivity of oxygen has long been recognized and discussed, e.g., in terms of the regulation of local blood flow, but the mechanism by which contraction is enhanced with increased oxygen partial pressure is still unknown (Sparks 1980, for review; Proctor and Duling 1982; Coburn et al. 1986). Additionally, in vitro experiments have shown that low molecular hydroperoxides (Heinle 1984) or fatty acid hydroperoxides (Sasaki et al. 1981) are able to induce vasoconstriction in relaxed arterial rings. On the other hand, the discovery of endothelial-mediated smooth muscle relaxation (Furchgott and Zawadzki 1980) has led to a number of studies showing that in precontracted vessel segments, reactive oxygen metabolites produce relaxation (Rubanyi and Vanhoutte 1986a; Coburn et al. 1986), which at least in some cases seems to be mediated by endothelium-dependent mechanisms. This means that the findings regarding the effects of oxygen and its metabolites on the function of arterial smooth muscle are still very controversial. The present study summarizes experiments which were performed in order to improve knowledge of the consequences of oxidative attack on metabolism and contraction of arterial smooth muscle.

Methods

The methods of the experiments cited here have already been described elsewhere. In brief, segments of carotid artery and thoracic aorta were obtained from male New Zealand rabbits (2.0–3.0 kg body weight). The aortic rings were used for metabolic studies. In these experiments, the tissues were incubated in the presence of hydroperoxides (H_2O_2 or tert-butylhydroperoxide, tBHP) and afterwards analyzed for either the degree of glutathione reduction (Heinle 1979a), the activity of glycogen phosphorylase a and b (Heinle 1982), or the glycogen content (Heinle 1985). The results were compared with those obtained with peroxide-free incubated tissues. Studies concerning the influence of hydroperoxides on smooth muscle contraction were performed with segments of carotid artery. Handling of the samples, the perfusion chamber and measuring device for recording isometric contraction have been described (Heinle 1984; Lindner and Heinle 1987). Stimulation was achieved by superfusing the arterial rings with hydroperoxides (H_2O_2, tBHP) in a concentration range of 0.1–10 mM. For comparison, KCl (60 mM), noradrenaline (10^{-6} M), and caffeine (10 mM) were used as further agonists. The influence of the oxygen partial pressure on the cellular thiol content was determined by incubating rabbit aortic segments either air equilibrated or under an atmosphere of pure N_2 or pure O_2 (at normal pressure). Afterwards, the tissues were homogenized, and the thiol content was determined with dithionitrobenzoic acid (Jocelyn 1972).

Results

Influence of Hydroperoxides on the Arterial Glutathione Redox State

The alterations in the aortic glutathione redox state induced by incubation in the presence of tBPH are shown in Fig. 1. Under the influence of peroxide, there is a strong increase in the glutathione disulfide/glutathione quotient, whereas total soluble glutathione is slightly decreased. The latter phenomenon may be due to formation of mixed disulfides with proteins which are not detectable in the glutathione assay used. Changing the incubation medium to peroxide-free solution

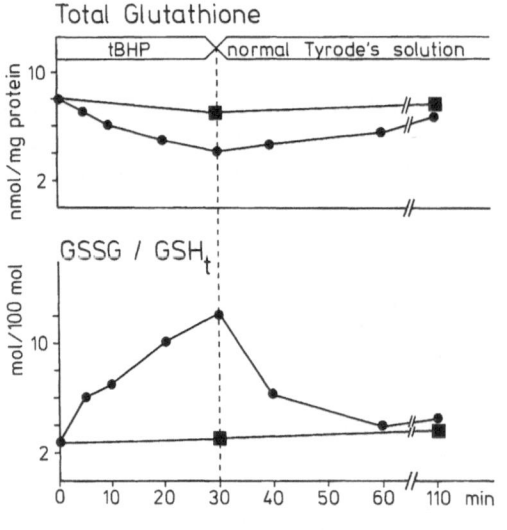

Fig. 1. Peroxide-induced alterations in content of total glutathione (GSSG + GSH) are shown *above* and degree of glutathione oxidation (GSSG/GSH$_t$) in segments of rabbit thoracic aorta *below*. tBHP (1.6 mM) was applied for 30 min in Tyrode's solution, after which the segments were incubated in peroxide-free solution. Controls were suspended in Tyrode's solution alone. This experiment was performed with segments of four rabbit aortas; a typical result from one aorta is presented

causes a slow normalization of the glutathione redox quotient as well as of the content of soluble glutathione. This means that tBHP is very effective in disturbing the cellular thiol/disulfide metabolism. However, these alterations seem to be completely reversible under the experimental conditions provided here.

Influence of Hydroperoxides on Arterial Glycogen Metabolism

As shown in Table 1, the incubation of aortic rings in the presence of hydroperoxides leads to a strong decrease in the glycogen content when the incubation medium does not contain glucose. The administration of hexose prevents this effect, indicating that the protecting systems of smooth muscle cells depend on glucose supply. In this context it seems that glucose is not only important as an energy-providing substrate but also in maintaining the $NADPH_2/NADP$ quotient via the pentose-phosphate cycle. In addition, the results show that peroxide-induced glycogen degradation is also achievable in a Ca^{2+}-free solution. This means that extracellular Ca^{2+} is not involved in glycogen degradation stimulated under these conditions.

Similar conclusions can be drawn from the experiments in which the peroxide-induced activation of glycogen phosphorylase a was measured. In glucose-free solution (Table 2), hydroperoxides cause a six fold increase in the ratio of phosphorylase a to total phosphorylase. The same effect is observed when the peroxide incubation was performed in Ca^{2+}-free solution. However, activation of phosphorylase a was diminished by the addition of glucose. This means that the observed alterations in aortic glycogen content correlate very well with the alterations in the activity of phosphorylase a. Additionally, activation of phosphorylase a is also achieved by the glutathione oxidizing agent diamide.

The increase in phosphorylase a might be the result of either an activation of the Ca^{2+}-dependent phosphorylase b kinase or of an inhibition of phosphorylase a phosphatase.

Table 1. Influence of H_2O_2 and tBHP on glucose and glycogen content in segments of rabbit thoracic aorta. Generally, the segments were incubated for 30 min. The results are given in µmol/g dry weight and represent mean values ± SD of experiments with segments of 6 rabbit aortas

	Glucose	Glycogen
Immediately after preparation	1.2 ± 0.6	14.3 ± 3.5
Incubation in Tyrode's solution		
without glucose	0.7 ± 0.5	11.5 ± 2.7
+ tBHP (1.6 mM)	0.8 ± 0.6	5.6 ± 3.5*
+ H_2O_2 (1.8 mM)	0.7 ± 0.5	4.2 ± 2.5*
Incubation with glucose (5 mM)	5.5 ± 1.3	17.2 ± 3.2
+ tBHP (1.6 mM)	4.6 ± 0.9	12.2 ± 2.2
+ H_2O_2 (1.8 mM)	5.2 ± 1.2	10.2 ± 3.0
Incubation in 0.9% NaCl + tBHP (1.6 mM)[a]	0.9 ± 0.3	5.5 ± 2.8

* Decrease in glycogen is significant with $P < 0.05$ (Wilcoxon-Wilcox)
[a] 2 rabbit aortas

Table 2. Influence of oxidative stress on the ratio of phosphorylase a to total phosphorylase in segments of rabbit thoracic aorta. Generally, the segments were incubated for 30 min. Phosphorylase a was determined without AMP, and total phosphorylase in the presence of 2 mM AMP. The results represent mean values ± SD of experiments with segments of 6 rabbit aortas

	Phosphorylase a/total phosphorylase (%)
Immediately after preparation	4.4 ± 2.8
Incubation in Tyrode's solution	
Without glucose	4.6 ± 3.0
+tBHP (1.6 mM)	27.2 ± 11.5
+ diamide (5 mM)[a]	24.5 ± 10.8
With glucose (5 mM)	5.0 ± 1.8
+tBHP (1.6 mM)[a]	6.9 ± 3.5
Incubation in 0.9% NaCl	
+tBHP (1.6 mM)[a]	32.6 ± 9.7

[a] 3 rabbit aortas

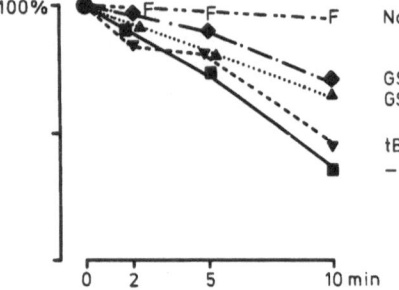

Fig. 2. Influence of glutathione disulfide (GSSG) on inactivation of glycogen phosphorylase a added to aortic homogenate. The agents administered to the assays are indicated

Since generally protein phosphatases need thiol protection to maintain activity in vitro, perhaps these enzymes in the cells are regulated by the thiol redox quotient. In order to evaluate the importance of this possible explanation for the peroxide-induced effects on glycogen phosphorylase a, the isolated enzyme was incubated with aortic homogenate completed with oxidized glutathione. The results are given in Fig. 2, showing that activity of the endogenous phosphatase very rapidly inactivates the added phosphorylase a. The known inhibitor of the phosphatase, NaF, blocks this inactivation completely, tBHP seems to be ineffectual, whereas glutathione disulfide causes a moderate effect. Although the concentration of the disulfide used (5 mM) seems to be unphysiologically high, it is conceivable that small alterations in the cellular thiol redox state might be involved in the regulation of the protein phosphatase, leading to detectable effects over a longer period of time.

Influence of Hydroperoxides on Contraction

To determine whether hydroperoxides influence Ca^{2+} metabolism in vascular smooth muscle, isometric contraction was used as a parameter indicating intracellularly increased levels of Ca^{2+}. The experiments show that hydroperoxides applied in concentrations higher than $2-4 \times 10^{-4}$ M produce contraction in segments of rabbit carotid artery. The process of contraction is usually characterized by a slow increase in force, the maximum being reached 5–10 min after withdrawal of the peroxide. After that, the muscle relaxes down to the initial tension. The dose dependency as obtained from rabbit carotid artery is given in Fig. 3. It shows that H_2O_2 is somewhat more efficient than tBHP, probably due to the lower molecular weight and higher diffusional transport.

Again, the protective effect of glucose is seen in Fig. 4. Reducing the concentration of glucose in the medium, the time needed for relaxation increases. In the absence of the carbohydrate, no relaxation occurred within the observation period of 30 min. Yet, to find this effect it is necessary to preincubate the vessel segment with the glucose-free solution for at least 20 min, indicating the importance of the clearance of intracellular stores. Similar to glycogen degradation, contraction is also induced in Ca^{2+}-free solution and furthermore by the thiol-oxidating agent diamide (Table 3). In Ca^{2+}-free solution there is a rapid decline of contractions induced by KCl, caffeine, or noradrenaline, whereas tBHP produced contractions reaching the same maximum force even during the second repetition (not shown in the table). This means that the stimulation is independent of extracellular Ca^{2+}. Nevertheless, it might depend on a Ca^{2+} signal related to liberation of Ca^{2+} from intracellular stores, which are different from those activated by caffeine and noradrenaline.

Diamide concentrations above 0.5 mM produced contraction in rings of rabbit carotid artery which were reversible upon withdrawal of the agent. This favors the conception that the degree of thiol oxidation is an important mediator of peroxide-induced effects on vascular smooth muscle.

Due to the assumption that in smooth muscle Ca^{2+} induces phosphorylation of myosin light chains which produces contraction via rapid cross-bridge cycling

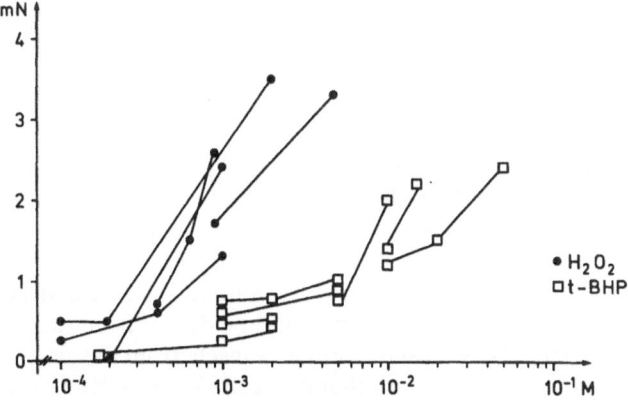

Fig. 3. Dose dependency of peroxide-induced contraction in segments of rabbit carotid artery. Values obtained from the same animal are connected

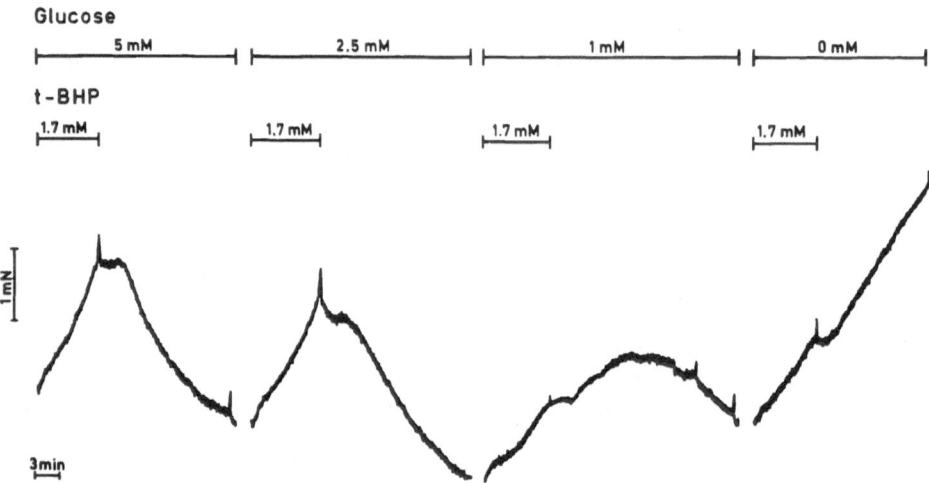

Fig. 4. Influence of glucose withdrawal on tBHP-induced contraction in rabbit carotid artery. The concentration of glucose and hydroperoxide are given at the *bars*. Prior to the application of tBHP, the decreased glucose concentration was perfused for at least 20 min

Table 3. Influence of Ca^{2+} on contraction force evoked by various stimuli in rings of rabbit carotid artery. The contractions were stimulated first by application of the agonist in normal Tyrode's solution (Ca^{2+} 2.5 mM). Then the measuring chamber was perfused with a modified solution in which Ca^{2+} was omitted and EGTA (4 mM) was added. In this solution, the same stimuli were applied. Results of a typical experiment

Stimuli	Contraction force (mN)	
	2.5 mM Ca^{2+}	0 mM Ca^{2+}
KCl (60 mM)	42	3.0
Noradrenaline (10^{-6} M)	16.2	4.3
Caffeine (10 mM)	8.5	2.7
H_2O_2 (1.7 mM)	3.4	4.8
Diamide (2 mM)	2.5	N.D.

N. D., not determined

with actin, we performed quick-release experiments with arterial rings stimulated by hydroperoxides. In principle, this method allows determination of the recovery of force in muscles contracting isometrically after a defined small, rapid shortening of muscle length. This gives an indication of the rate of the cross-bridges.

However, the passive mechanic properties of vascular smooth muscle conflict with the analysis of the results, so that no conclusion could be drawn from the experiments with arterial rings. Therefore, we used papillary muscle from the rat and compared the peroxide-induced contraction with that induced either by KCl or by hypoxia (Holubarsch et al. 1985). The results clearly demonstrate an active recovery of force during H_2O_2 stimulation, which is about 80% of that obtained during potassium-induced depolarization. In hypoxia, no recovery is found, indicating the rigidity of the myosin-actin bridges. This means that in muscles of the rat heart, H_2O_2 evokes a release of Ca^{2+}, which then produces contraction with rapid-cycling myosin-actin bridges.

Influence of Oxygen on Arterial Thiol Content

The alterations in thiol content detectable with dithionitrobenzoic acid in aortic tissue incubated either under air equilibration, or under oxygen or nitrogen (normal pressure) are given in Table 4. Values obtained with vessel segments incubated in the presence of H_2O_2 are also included. As expected, the highest thiol content is found under nitrogen atmosphere; the lowest after incubation in the presence of the peroxide. Oxygen atmosphere produces a medium decrease in the thiol content, which, interestingly enough, is not significantly different from that obtained under air equilibration. This means that oxygen has an influence on the cellular content of reduced thiol groups with susceptibility depending on a partial pressure of oxygen between 0 and 150 torr. Therefore, one can conclude that the thiol redox system is sensitive to changes in oxygen partial pressure occurring in vivo. However, the question whether these oxygen-induced alterations in the thiol system produce effects similar to those induced under the influence of hydroperoxides cannot be answered at the moment.

Preliminary results show that in our mechanic system no alteration in vessel wall tension is found when the gas equilibration is changed from N_2 to O_2 and vice versa. However, when potassium depolarization-induced contraction is measured either under N_2 or under O_2, usually a higher contraction force is seen in the latter case. This result can be discussed in terms of a facilitated Ca^{2+} liberation under oxygenated conditions. However, one should be careful in this interpretation since oxygen deprivation has also far-reaching consequences on energy metabolism and thus on all energy-dependent functions.

Table 4. Influence of oxygenation on thiol content of segments of rabbit thoracic aorta. The segments of 5 aortas were incubated for 30 min under the indicated conditions. Thiol content was determined by dithionitrobenzoic acid. The values are given in relation to that found under nitrogen ($=100\%$). During nitrogen bubbling, the PO_2 decreased to about 2–3 torr

Incubation condition	Relative thiol content (%)
Nitrogen	100
Air	74.5 ± 4.2
Oxygen	73.5 ± 5.6
H_2O_2 (1 mM)	50.7 ± 14.9

Discussion

Mechanisms by Which Hydroperoxides Might Influence Glycogen Metabolism in Arterial Smooth Muscle

The enzyme cascade regulating glycogen degradation is one of the best-known in cellular metabolism, and the last steps – the regulation of the activity of glycogen phosphorylase – are outlined schematically in Fig. 5. The inactive phosphorylase b is phosphorylated by phosphorylase b kinase, thus forming the activated phorphorylase a. Inactivation occurs via dephosphorylation by a protein phosphatase. Phosphorylase b kinase itself is the target of complex regulatory mechanisms initiated by Ca^{2+} or cAMP. Although they are not less important, little is known about the protein phosphatases. In liver it was shown that the enzyme responsible for inactivation of glycogen phosphorylase a is inhibited by glutathione disulfide (Usami et al. 1980). Therefore, the question arises whether glycogen degradation induced under oxidative stress as observed in vascular smooth muscle (Table 1) and liver (Sies et al. 1974) is mediated via Ca^{2+} mobilization, activating the kinase pathway, or via increase of GSSG, inhibiting phosphorylase a phosphatase.

Although glycogen degradation was also found in Ca^{2+}-free incubation solution, intracellular Ca^{2+} liberation might be an important signal in transducing the metabolic effects of oxidative stress. This point will be further discussed in connection with the contraction-inducing effects.

Here, we shall focus on alterations in thiol redox state and possible influences on protein phosphatases. As shown, oxidative stress induces a decrease in thiol content and an increase of glutathione disulfide, which is caused either by spontaneous oxidation or by detoxification through glutathione peroxidase. However, complete detoxification is not achieved until GSSG is reduced by glutathione disulfide reductase. Glutathione peroxidase and glutathione disulfide reductase were shown to be present in arterial smooth muscle in a 3:1 ratio of their specific activities (Heinle 1979b). This means that the reductase is a limiting enzyme which connects the degree of thiol reduction to the supply of NADPH-related reducing equivalents emerging from the pentose phosphate cycle (Fig. 5). An increased turnover through the pentose phosphate cycle under the influence of hydroperoxides was measured in various cellular systems (for review see Chance et al. 1979). This relationship provides a plausible explanation for the observed protection by glucose against the peroxide-induced metabolic effects. In the absence of extracellular glucose, glycogen is degraded, providing the necessary carbohydrate monomers. Whether inhibition of the protein phosphatase by thiol oxidation is involved in the regulation of glycogen breakdown under physiological conditions is unknown. Nevertheless, the results presented here show that oxidized glutathione – yet at higher concentrations than found intracellularly – exerts a remarkable inhibition on smooth-muscle phosphorylase phosphatase. However, an increased ratio of phosphorylase a/b cannot be the result of inhibition of the phosphatase alone. There must either be a basic turnover of phosphorylase b to phosphorylase a without Ca^{2+} stimulation, or there is concomitant peroxide-induced liberation of Ca^{2+} activating the kinase pathway. Only then can a shift to activation of phosphorylase a be expected.

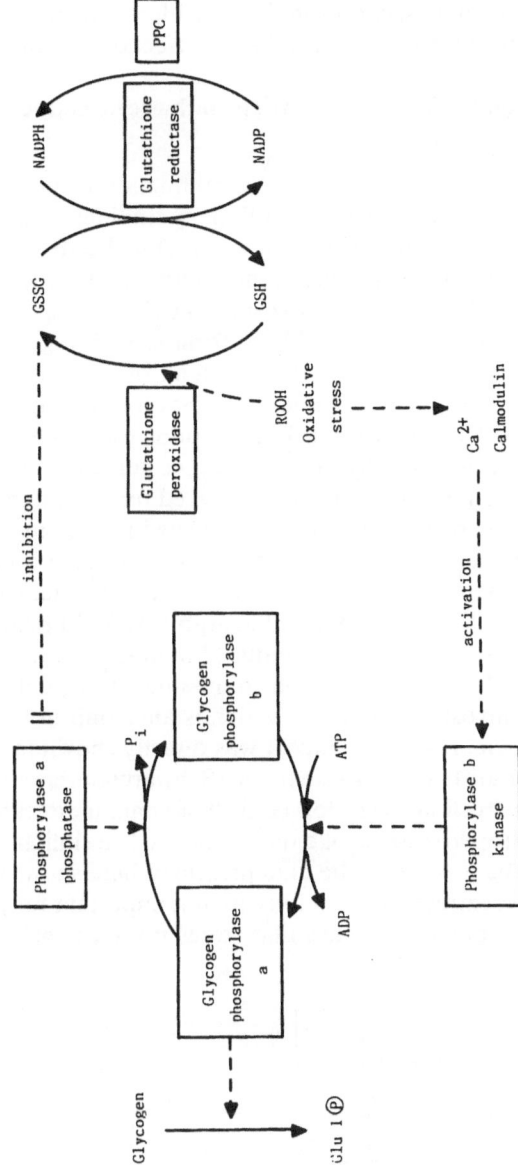

Fig. 5. Reaction scheme indicating possible pathways by which oxidative stress could influence glycogen degradation

As discussed in the next section, similar considerations must be taken into account with relation to the contraction-inducing effect of hydroperoxides.

Mechanisms by Which Hydroperoxides Might Influence Contraction of Arterial Smooth Muscle

Although the mechanism of smooth muscle contraction is the subject of an overwhelming amount of research, there are still many findings which are more or less merely hypothetically explained (for review see Hartshorne and Gorecka 1980). As shown in Fig. 6, it is widely accepted that contraction is induced by an increase of cytoplasmic Ca^{2+} which activates myosin light-chain kinase. After phosphorylation of the light chains, myosin is able to form cross-bridges with actin, which themselves are independent of Ca^{2+} (Gagelmann et al. 1984; Kamm and Stull 1985). Relaxation will occur after dephosphorylation of the light chains by a protein phosphatase and Ca^{2+} extrusion. Explanations for maintained smooth muscle contractions are rather sketchy. Some evidence suggests that an increased degree of phosphorylation of myosin is only achieved for several minutes. After that, tension can still be generated with dephosphorylated myosin in a Ca^{2+}-dependent mechanism. However, this mechanism has slow cross-bridge cycling, whereas the interaction between phosphorylated myosin and actin is characterized by a high rate of cross-bridge cycling (Rembold and Murphy 1986). In relation to the discussion in the preceding section, peroxide-induced inhibition of the protein phosphatase alone cannot explain the contraction-stimulating effect, unless there is a basic phosphorylation of myosin light chains which is antagonized by the phosphatase in unstimulated muscle. However, since it was directly shown in liver cells (Sies et al. 1981; Orrenius et al. 1983), adipose tissue (Schwartz-Sørensen et al. 1980), and, recently, in a cultured cell line (Hyslop et al. 1986) and, indirectly, in heart muscle (see above) that hydroperoxides evoke release of Ca^{2+}, a similar effect on vascular smooth muscle can be expected. The data presented here show that, at least in arterial smooth muscle, extracellular Ca ions are not important for peroxide-induced contraction, and that putative intracellular stores do not relate to those activated

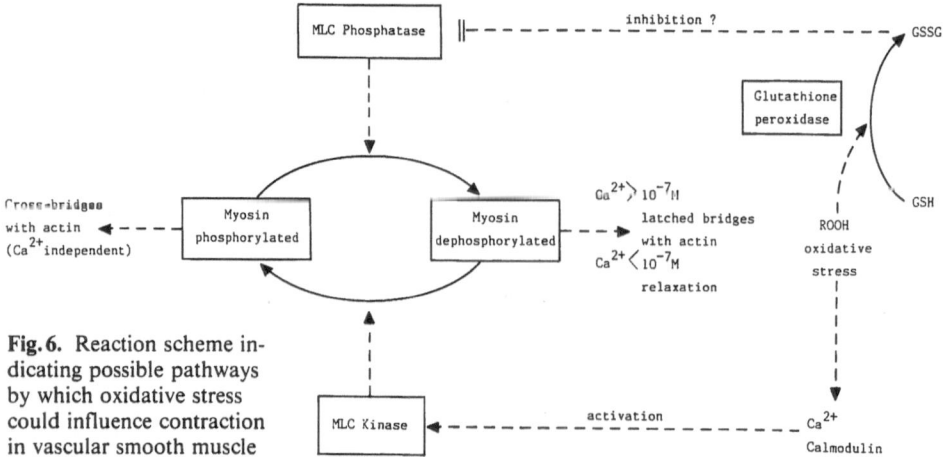

Fig. 6. Reaction scheme indicating possible pathways by which oxidative stress could influence contraction in vascular smooth muscle

by noradrenaline or caffeine. Interestingly enough, several groups have shown that mitochondria (Lötscher et al. 1979; Orrenius et al. 1983) and endoplasmatic reticulum (Trimm et al. 1986) are able to release Ca^{2+} upon oxidation of thiol groups and - in mitochondria - of nicotinamide nucleotides. Thus, one can conclude that the degree of thiol oxidation is not only a regulator for protein phosphatases but also a strong signal for intracellular Ca^{2+} release.

This Ca^{2+} liberation could be responsible for the contraction stimulating effect of hydroperoxides observed in relaxed rings of carotid artery (see above). Similarly, contractions were found in rings of other peripheral arteries and veins (Kling and Heinle 1986), in rings of cerebral arteries (Sasaki et al. 1981), and of bronchial smooth muscle (Stewart et al. 1981).

On the other hand, there are some further observations on peroxide or free radical effects on vascular smooth muscle which conflict with this interpretation. It was found that arterial rings precontracted with $PGF_{2\alpha}$ relax after addition of hydroperoxide. Since a concomitant increase in cellular cGMP was observed, this effect was related to the action of an endothelium-derived relaxing factor (Rubanyi and Vanhoutte 1986a; Thomas and Ramwell 1986).

However, as the mechanism of $PGF_{2\alpha}$ stimulation in vascular smooth muscle is still unknown (Bolton 1979) and the mechanism of maintained contraction obscure, the explanations for the findings are rather hypothetical. Although a mechanism of cGMP-induced smooth muscle relaxation has been proposed (Rapoport 1986), contradicting results have been published which describe relaxation with decreased cGMP (Coburn et al. 1986). Therefore, one cannot exclude the possibility that oxidative stress has different effects on smooth muscle, depending on its contractile state, or that hydroperoxides have differential effects on vascular tone when acting primarily on the endothelial lining.

Implications for a Thiol-Dependent Oxygen Sensor

Possible mechanisms by which the tone of vascular smooth muscle is regulated in dependency of PO_2 have been discussed by Sparks (1980). On the one hand, one must note that there exist large differences in the susceptibility to oxygen in various parts of the vascular tree, and that pulmonary vessels have the contradictory property of hypoxia-induced contraction. On the other hand, it has been shown that, in many cases, relaxation could not be explained by decreased availability of ATP due to limited oxygen supply.

Thus, it was discussed by Rubanyi and Paul (1985) that in porcine coronary artery, at an oxygen pressure of about 70–300 torr, prostaglandin metabolism is influenced, leading in this case to vasoconstriction. At an oxygen pressure near zero, a limited energy supply should be responsible for relaxation. Further observations suggest that reactive oxygen species may inactivate the endothelium-derived relaxing factor (Rubanyi and Vanhoutte 1986b), and that, in some parts of the vascular tree, the oxygen regulation of vascular tone is mediated by the endothelium, but not in others (Coburn et al. 1986). These authors suggest that cellular cGMP levels are influenced by oxygen, but are not involved in hypoxia-induced relaxation. This means that clear mechanisms by which tension in vascular smooth muscle could be regulated by O_2 cannot be postulated at the moment.

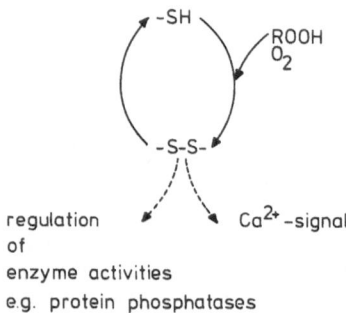

regulation
of
enzyme activities
e.g. protein phosphatases

Fig. 7. Simplified model indicating the possible importance of the cellular disulfide/thiol quotient as a metabolic mediator of oxidative stress

As seen from the results presented here, the cellular degree of thiol oxidation is influenced by the partial pressure of oxygen within a range of 0–150 torr. It is not known whether this effect is mediated by spontaneous oxidation of SH groups, by oxygen metabolites produced, e.g., by mitochondria in dependency of PO_2 or by oxygen-dependent enzymes. Furthermore, it is unclear whether the alterations in the thiol groups include those whose function seems to be related to the effects of hydroperoxides on glycogen metabolism and contraction.

Nevertheless, the results of preliminary experiments show that depolarization induces contractions whose maximum force is higher in the presence of oxygen than in its absence. Similar findings have been obtained in other in vitro systems (Sparks 1980; Christensen et al. 1982). Therefore, in Fig. 7 a very simplified mechanism is proposed which links the oxidation potential of oxygen and oxidative stress to the degree of oxidation of functional thiol groups. These might be involved in further signal transduction, leading to alterations in cellular Ca^{2+} homeostasis and to alterations in the activity of SH-dependent enzymes. These results are in agreement with the concept of blood-flow regulation, suggesting that a sufficient oxygen supply in the peripheral arterial system causes vasoconstriction and hypoxia vasodilatation. Further work is necessary to substantiate the physiological role of this proposed mechanism in chemoception of the partial pressure of oxygen.

Acknowledgements. Parts of this work were supported by the Deutsche Forschungsgemeinschaft (He 1083). Excellent technical assistance was provided by Ms. H. Ableiter and Ms. R. Vesenmaier. Thanks are extended to Ms. A. Roth for typing the manuscript

References

Bolton TB (1979) Mechanisms of action of transmitters and other substances on smooth muscle. Physiol Rev 59 606–718

Chance B, Sies H, Boveries A (1979) Hydroperoxide metabolism in mammalian organs. Physiol Rev 59: 527–605

Christensen HJ, Østgaard SE, Andreasen G (1982) The influence of PO_2, pH, and albumin on the in vitro contraction of vascular smooth muscle. J Pharmacol Method 8: 99–108

Coburn RF, Eppinger R, Scott DP (1986) Oxygen-dependent tension in vascular smooth muscle. Circ Res 58: 341–347

Del Maestro RF (1980) An approach to free radicals in medicine and biology. Acta Physiol Scand [Suppl] 492: 153-168

Fridovich I (1983) Superoxide radical: an endogenous toxicant. Annu Rev Pharmacol Toxicol 23: 239-257

Furchgott RF, Zawadzki JV (1980) The obligatory role of endothelial cells in the relaxation of arterial smooth muscle by acetylcholine. Nature 288: 373-376

Gagelmann M, Mrwa U, Bostrum S, Rüegg JC, Hartshorne D (1984) Effect of Ca^{2+}-independent myosin light chain kinase on different skinned smooth muscle fibers. Pflugers Arch 401: 107-109

Halliwell B (1978) Biochemical mechanisms accounting for the toxic action of oxygen on living organisms: the key role of superoxide dismutase. Cell Biol Int Rep 2: 113-128

Hartshorne DJ, Gorecka A (1980) Biochemistry of the contractile proteins of smooth muscle. In: Bohr DF et al. (eds) Vascular smooth muscle. Am Physiol Soc, Bethesda, pp 93-120 (Handbook of physiology, sect II, vol II)

Heinle H (1979a) The degree of glutathione oxidation in excised aortic tissues of rats and rabbits. Hoppe-Seyler's Z Physiol Chem 360: 1113-1116

Heinle H (1979b) The specific activities of glutathione peroxidase and glutathione reductase in homogenates of aortic segments of the rat. Hoppe-Seiler's Z Physiol Chem 360: 1157

Heinle H (1982) Peroxide-induced activation of glycogen phosphorylase a activity in vascular smooth muscle. Biochem Biophys Res Commun 107: 597-601

Heinle H (1984) Vasoconstrictions of carotid artery induced by hydroperoxides. Arch Int Physiol Biochim 92: 13-17

Heinle H (1985) Stoffwechseländerungen der Gefäßwand bei experimenteller Atherosklerose. Habilitationsschrift, Medizinische Fakultät, Tübingen

Holubarsch Ch, Takeda N, Heinle H (1985) Die H_2O_2-Kontraktur des Rattenmyokards - Vergleich mit Kaliumchlorid (KCl) und hypoxischer Kontraktur. Z Kardiol 74: 82

Hyslop PA, Hinshaw DB, Schraufstätter IU, Sklar LA, Spragg RG, Cochrane ChG (1986) Intracellular calcium homeostasis during hydrogen peroxide injury to cultured $P388D_1$ cells. J Cell Physiol 129: 356-366

Jocelyn PC (1972) Biochemistry of the SH group; the occurrence, chemical properties, metabolism and biological function of thiols and disulphides. Academic, London

Kamm KE, Stull ZT (1985) The function of myosin light chain kinase phosphorylation of smooth muscle. Annu Rev Pharmacol Toxicol 25: 593-620

Kling D, Heinle H (1986) Charakterisierung aktiver und passiver mechanischer Eigenschaften von Arterien und Venen. In: Hoffmeister HE (ed) Reagibilität der arteriellen und venösen Strombahn. Wissenschaftliche Verlagsgesellschaft, Stuttgart, pp 9-13

Lindner V, Heinle H (1987) Does the xanthine-xanthine oxidase system alter contractile behavior of vascular smooth muscle? Pflugers Arch 408: 204-206

Lötscher HR, Winterhalter KH, Carafoli E, Richter Ch (1979) Hydroperoxides can modulate the redox state of pyridine nucleotides and the calcium balance in rat liver mitochondria. Proc Natl Acad Sci USA 76: 4340-4344

Orrenius S, Jewell SA, Bellomo G, Thor H, Jones DP, Smith MT (1983) Regulation of calcium compartmentation in the hepatocyte - a critical role of glutathione. In: Larsson A et al. (eds) Functions of glutathione: biochemical, physiological, toxicological and clinical aspects. Raven, New York, pp 261-271

Proctor KG, Duling BR (1982) Oxygen-derived free radicals and local control of striated muscle blood flow. Microvasc Res 24: 77-86

Rapoport RM (1986) Cyclic guanosine monophosphate inhibition of contraction may be mediated through inhibition of phosphatidylinositol hydrolysis in rat aorta. Circ Res 58: 407-410

Rembold ChM, Murphy RA (1986) Myoplasmic calcium, myosin phosphorylation, and regulation of the cross-bridge cycle in swine arterial smooth muscle. Circ Res 58: 803-815

Rubanyi G, Paul RJ (1985) Two distinct effects of oxygen on vascular tone in isolated porcine coronary arteries. Circ Res 56: 1-10

Rubanyi GM, Vanhoutte PM (1986a) Oxygen-derived free radicals, endothelium, and responsiveness of vascular smooth muscle. Am J Physiol 250: H815-H821

Rubanyi GM, Vanhoutte PM (1986b) Superoxide anions and hyperoxia inactivate endothelium-derived relaxing factor. Am J Physiol 250: H822-H827

Sasaki T, Wakai S, Asano T, Watanabe T, Kirino T, Sano K (1981) The effect of a lipid hydroperoxide of arachidonic acid on the canine basilar artery. J Neurosurg 54: 357–365
Schwartz-Sørensen S, Christensen F, Clausen T (1980) The relationship between the transport of
glucose and cations across cell membranes in isolated tissues. Biochim Biophys Acta 602:
433–445
Sies H, Graf P, Estrela JM (1981) Hepatic calcium efflux during cytochrome P-450-dependent
drug oxidations at the endoplasmic reticulum in intact liver. Proc Natl Acad Sci USA 78:
3358–3362
Sies H, Gerstenecker C, Summer KH, Menzel H, Flohé L (1974) Glutathione-dependent hydroperoxide metabolism and associated metabolic transitions in hemoglobin-free perfused rat liver.
In: Flohé L et al. (eds) Glutathione. Thieme Stuttgart, pp 261–276
Sparks HV (1980) Effects of local metabolic factors on vascular smooth muscle. In: Bohr DF et
al. (eds) Vascular smooth muscle. Am Physiol Society, Bethesda, pp 475–513 (Handbook of
Physiology, Sect II vol II)
Stewart RM, Weir EK, Montgomery MR, Niewoehner DE (1981) Hydrogenperoxide contracts
airway smooth muscle: a possible endogenous mechanism. Respir Physiol 45: 333–342
Thomas G, Ramwell P (1986) Induction of vascular relaxation by hydroperoxides. Biochem Bio-
phys Res Commun 139: 102–108
Trimm JL, Salama G, Abramson JJ (1986) Sulfhydryl oxidation induces rapid calcium release
from sarcoplasmic reticulum vesicles. J Biol Chem 261: 16092–16099
Usami M, Matsushita H, Shimazu T (1980) Regulation of liver phosphorylase phosphatase by
glutathione disulfide. J Biol Chem 255: 1928–1931

Free Cytosolic Adenosine Sensitively Signals Myocardial Hypoxia

J. Schrader and A. Deussen

Department of Physiology I, University of Düsseldorf, Moorenstr. 5, 4000 Düsseldorf, FRG

Introduction

The first description of the coronary vasodilatory action of adenosine was published in 1929 by Drury and Szent-Györgyi. However, it was only in 1963 that Berne and independently Gerlach demonstrated that cardiac muscle can produce adenosine (Gerlach et al. 1963). This has led to the hypothesis that adenosine my be of major importance in the adjustment of coronary blood flow to the metabolic demands of the heart (Berne 1963). In the past 20 years numerous studies have tested the implications of the adenosine hypothesis (for review see Berne and Rubio 1979; Feigl 1983; Olsson and Bünger 1987). Although we have learned many important details concerning the site and mechanism of formation of adenosine, as well as of its mode of action (Schrader 1983; Sparks and Bardenheuer 1986), it is still a matter of debate to what extent this vasoactive nucleoside mediates changes in coronary blood flow under physiological conditions.

The purpose of this chapter is to review some of the recent evidence from our laboratory indicating that the formation of adenosine is closely linked to cardiac energy state. Aside from its proposed role in regulation of coronary blood flow, evidence will be provided suggesting that free cytosolic adenosine can serve as a sensitive index of cardiac hypoxia and ischemia.

Compartmentalization of Cardiac Adenosine Metabolism

Four sites of adenosine formation within the heart have been defined. As schematically outlined in Fig. 1, adenosine can be formed:

1. By the cardiomyocyte including the pathway from AMP (5'-nucleotidase) and S-adenosylhomocysteine (SAH-hydrolase). Under normoxic conditions hydrolysis of SAH prevailes, while during hypoxia the pathway from AMP appears to dominate (Lloyd and Schrader 1987). Adenosine is formed intracellularly and reaches the extracellular space by diffusion along its concentration gradient (Schütz et al. 1981; Schrader et al. 1981; Newby et al. 1983).
2. By vascular smooth muscle. These cells have been shown to constantly produce

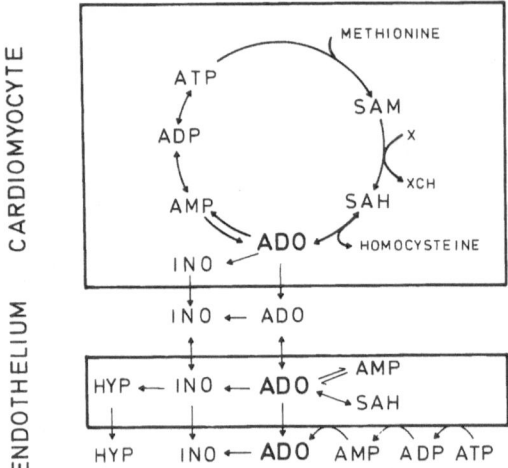

Fig. 1. Schematic representation of the different metabolic pathways and cellular sites of production of adenosine in the heart

adenosine, suggesting a direct influence on vascular smooth muscle tone (Belloni et al. 1986).

3. By the coronary endothelium which has been demonstrated to be equipped with all enzymes necessary for adenosine formation (Mistry and Drummond 1986). Endothelium-derived adenosine is continuously released into the coronary circulation (Deussen et al. 1986). Compared with the cardiomyocytes about 14% of the total cardiac adenosine release under normoxic conditions is derived from the coronary endothelium (Kroll et al. 1987). The relative contribution of this compartment decreases during tissue hypoxia and β-adrenergic stimulation of the heart (Deussen et al. 1986; Kroll et al. 1987).

4. By ecto 5'-nucleotidase localized on the luminal side of the vascular endothelium. Extracellular degradation of ATP is very rapid and little will escape in the coronary venous effluent as ATP (Gordon et al. 1986). The source for extracellular nucleotides is not well defined, but may include cardiomyocytes (Forrester and Williams 1977), sympathetic nerves (Fredholm et al. 1982), endothelial cells (Pearson et al. 1980), and platelets (Fukami et al. 1976).

The major site of adenosine formation during normoxia and hypoxia is the cardiomyocyte (Kroll et al. 1987), but other cells within the heart, such as smooth muscle and endothelial cells can also produce adenosine and may influence vascular tone directly. Due to the different sources for adenosine within the heart, diffusion distances to the site of adenosine action – adenosine receptors on vascular smooth muscle, coronary endothelium, and cardiomyocytes – must be assumed to be also different, and may profoundly influence the effective concentration of this nucleoside. Therefore compartmentalization of the metabolism of adenosine within the heart not only raises interesting questions concerning cell to cell communication via adenosine, but also illustrates the inherent difficulties when evaluating the physiological role of adenosine for cardiac function.

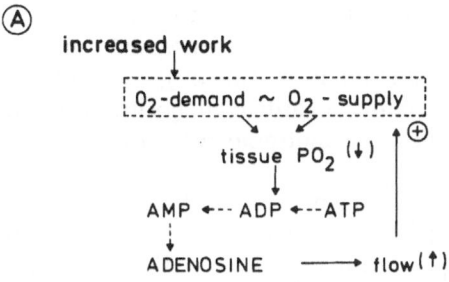

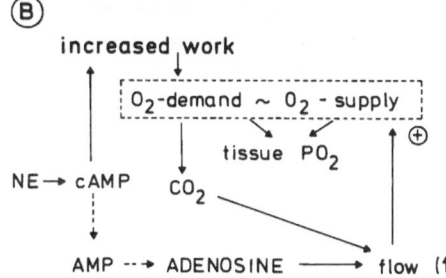

Fig. 2A, B. Possible mechanisms by which adenosine may be formed by the heart in order to maintain the ratio of oxygen supply to oxygen demand in equilibrium when work of the heart is increased

Signals for Adenosine Formation

In the original formulation of the adenosine hypothesis it was proposed, that whenever the oxygen demand of the heart is increased, adenosine is formed at an accelerated rate, which then causes the adaptive change in coronary blood flow (Berne 1963). In this model of metabolic coronary vasoregulation, adenosine was suggested to signal myocardial oxygen tension and to act in a feedback controlled system. It is the supply to demand ratio for oxygen which is the signal for an accelerated adenosine formation (Fig. 2A).

According to another model proposed later (Berne 1980), the production of adenosine is coupled to the rate of cardiac energy metabolism, adenosine being formed in direct proportion to the amount of ATP utilized (Fig. 2B). In this system, and from the standpoint of control theory, adenosine is not a primary feedback-controlled variable. Similarly to the mechanism of formation of CO_2, myocardial oxygen consumption would be the prime determinant of the rate of adenosine formation. In the case of stimulation of the heart with catecholamines, this may include the following hypothetical metabolic events (Fig. 2B). The catecholamine-induced increase in cyclic AMP not only triggers the changes in contractile force and thereby determines energy requirements, but also – via degradation of cyclic AMP to AMP and adenosine – forms a vasoactive metabolite by which the oxygen demand of the heart can be matched by a proportional increase in coronary blood flow. Thus, a single metabolic pathway could carry the information to signal changes in contractile force and oxygen supply to the heart in a proportional manner. As a secondary effect of this control mechanism tissue PO_2 would remain largely unchanged. Studies carried out in the isolated heart, however, make it rather unlikely that this latter pathway is the dominant mechanism

(Bardenheuer and Schrader 1986), since inotropic stimuli which do not act via cyclic AMP such as quabain and AR-L-115, did not cause less adenosine formation compared with β-adrenergic stimulation. In addition, increasing oxygen supply by overperfusion of the coronary system greatly reduced cardiac adenosine efflux in the presence of isoproterenol without altering the inotropic response (Bardenheuer and Schrader 1986).

In order to evaluate the above-outlined two hypotheses, we have carried out experiments in an isolated working guinea pig heart preparation. As is shown in Fig. 3, β-adrenergic stimulation with isoproterenol increased the efflux of adenosine which perfectly paralleled the changes in coronary flow and MVO_2. A completely different picture was obtained when oxygen demand of the heart was elevated by increasing the afterload. Clearly, adenosine release did not increase even though MVO_2 was changed to a similar extent compared with isoproterenol. The

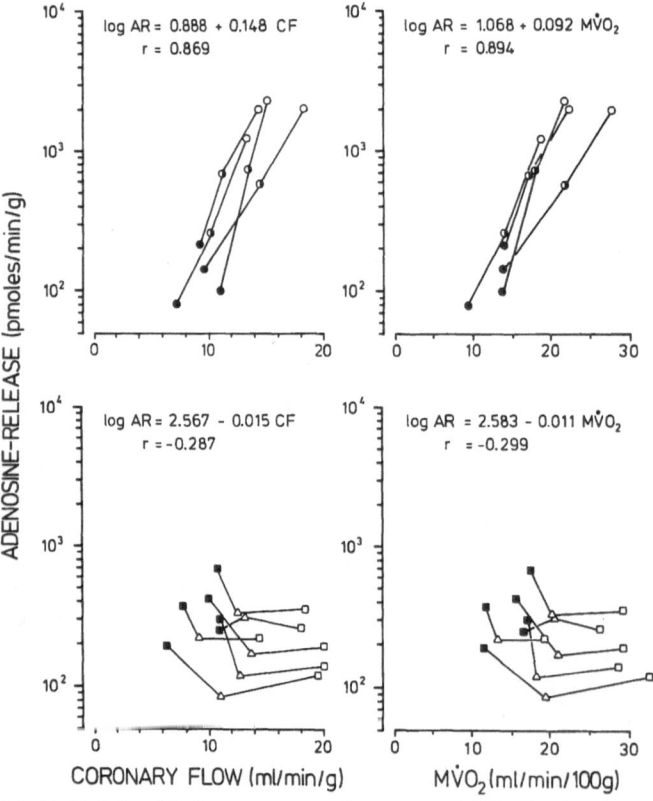

Fig. 3. Relationship between adenosine release *(AR)*, coronary flow *(CF)*, and myocardial oxygen consumption *(MVO₂)* after infusion of isoproterenol and following changes of afterload. Data given in *top panel* represent values obtained without (●), and with 1.5×10^{-9} (◐) and 6.0×10^{-9} (○) isoproterenol (afterload, 50 mmHg; coronary perfusion pressure, 50 mmHg; preload, 7.4 mmHg). Values given in *bottom panel* were obtained by increasing afterload and perfusion pressure from 40 (■) to 60 (△) and 80 mmHg (□) (preload, 7.4 mmHg) (From Bardenheuer and Schrader 1983)

most likely explanation of these experiments is that increasing the afterload also increases coronary perfusion pressure and thereby oxygen supply to the heart. In accordance with the first hypothesis (Fig. 2 A), this readjusts tissue PO_2, thereby removing the signal for adenosine formation. In fact, the measured increase in coronary flow was pressure dependent since calculated coronary resistance did not change (Bardenheuer and Schrader 1983). These and further experiments in which the supply to demand ratio for oxygen was systematically varied (Bardenheuer and Schrader 1986) suggested that not the metabolic rate as such is the major stimulus for the formation of adenosine, but the imbalance between oxygen delivery and oxygen demand.

Relating cytosolic adenylates to the formation of adenosine, Bünger and Soboll (1986) have suggested that the adenosine production by the heart may solely be regulated by the free cytosolic AMP concentration through the ATP potential. Similarly Nuutinen et al. (1984) proposed that coronary blood flow in the isolated heart is inversely proportional to the cytoplasmatic phosphorylation potential. In this model, the activity of myokinase would translate changes in the ATP/ADP ratio into changes in free AMP (Bünger and Soboll 1986). The increased availability of ADP not only would constitute the driving force for the formation of AMP and adenosine, but would also signal the mitochondria to increase oxidative phosphorylation (Jacobus 1985). It must be emphasized, however, that this concept is mainly based on studies performed on isolated hearts perfused with a saline medium. The oxygen-carrying capacity of a saline medium is low and there are reasons to assume that oxygenation of cardiac muscle even in the unstressed heart may be borderline (Becker and Gerlach 1987). Particularly in situations when the oxygen demand is increased, tissue PO_2 of saline-perfused hearts is expected to fall more steeply than in the blood-perfused heart. Consequently, changes in the supply to demand ratio for oxygen, triggering the formation of adenosine, may be more pronounced in the buffer-perfused heart than in the blood-perfused heart. It may well be that in a oxygenated heart in situ the range in which tissue PO_2 is maintained constant is broader and that mechanisms other than adenosine are available to adjust coronary flow and oxydative phosphorylation to increasing metabolic demands in this regulatory range. In fact, it has recently been shown with [31]-P-NMR-spectroscopy that free ADP in the dog heart in situ did not significantly change despite the fact that work performed by the heart was increased by atrial pacing (Balaban et al. 1986). Similarly, changes in coronary flow elicited by arterial pacing were not associated by changes in the formation of adenosine (Manfredi and Sparks 1982). Concerning the mechanism responsible for the observed changes in metabolic rate in view of unchanged free ADP, an alternative hypothesis was proposed by McCormack and Denton (1986). The increase in the cytoplasmic concentration of Ca^{2+} during increased cardiac work not only triggers changes in ATPase activity, it can also stimulate rates of pyruvate oxidation and citrate cycle flux, thereby increasing the ATP supply to meet the enhanced demand (McCormack and Denton 1986).

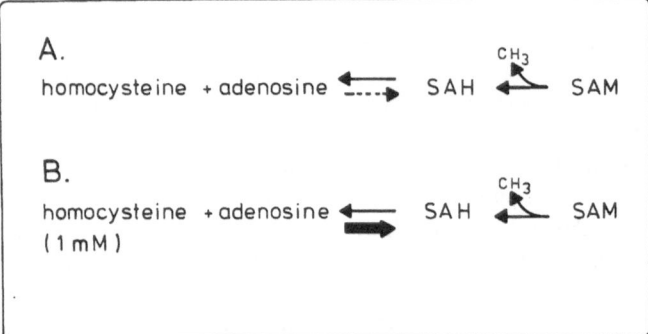

Fig. 4. A Under control conditions S-adenosylhomocysteine (SAH) formed via the transmethylation pathway from S-adenosylmethionine (SAM) is degraded to adenosine and homocysteine, since both adenosine and homocysteine are effectively further metabolized. **B** In the presence of saturating concentrations of homocysteine net synthesis of SAH from adenosine and homocysteine prevails, free cytosolic adenosine becoming the rate-limiting factor for SAH synthesis.

S-Adenosylhomocysteine and Free Cytosolic Adenosine

Two fractions of adenosine in cardiac muscle must be differentiated: a large protein-bound pool which is biologically inactive and free cytosolic adenosine which constitutes the biologically active fraction. Several lines of evidence suggest that the bulk of the myocardial adenosine is bound to S-adenosylhomocysteine hydrolase (Ueland 1982), a cytosolic enzyme. The adenosine-SAH-hydrolase complex is rather stable; the half-life of adenosine binding in the dog heart was determined to be several hours (Olsson et al. 1982). Measuring total adenosine content of cardiac muscle by conventional tissue extraction procedures may therefore miss small changes in the free adenosine concentration due to the large "background" of bound adenosine. At present there is no differential extraction procedure available which permits the determination of the free adenosine concentration in cardiac tissue directly.

In the following in new approach will be described for the accurate and sensitive assessment of changes in free adenosine within the heart. Crucial for the understanding of this method are the kinetics of SAH-hydrolase. This enzyme not only tightly binds adenosine, it also is capable of producing adenosine from SAH which itself is derived from cellular SAM-dependent methylation reactions (Ueland 1982). Although the equilibrium constant of SAH-hydrolase favors SAH synthesis (De la Haba and Cantoni 1959), the reaction normally proceeds in the direction of hydrolysis because the products of this reaction, adenosine and l-homocysteine, are rapidly removed (Fig. 4A). However, under conditions of increased levels of adenosine and l-homocysteine, the reaction is reversed and SAH is produced (Fig. 4B). SAH constitutes the endproduct of this reaction and accumulates intracellularly without being further metabolized. In the presence of saturating concentrations of l-homocysteine ($> 200\ \mu M$), the rate-limiting step in the formation of SAH is solely adenosine. Enzyme kinetics predict that under these conditions the rate of SAH accumulation is directly proportional to the concentra-

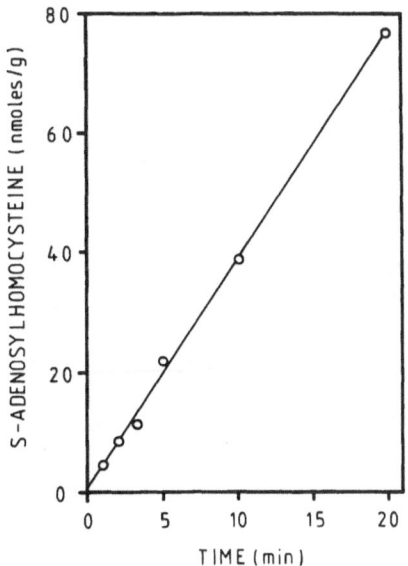

Fig. 5. Effect of intracoronary infusion of homocysteine (1 mM) and adenosine (10 μM) on SAH tissue levels of isolated perfused nonworking guinea pig hearts. Hearts were paced at 300/min and stop-frozen after 1, 2, 3, 5, 10, and 20 min of homocysteine infusion

tion of free adenosine. Since SAH-hydrolase is a cytosolic enzyme, it senses changes of adenosine in this cellular compartment.

In an attempt to validate the basic assumptions of the above-outlined concept, experiments were performed in the isolated guinea pig heart perfused with a modified Krebs-Henseleit medium (Deussen et al. 1986). Basal values of SAH determined with HPLC techniques were 1.1 ± 0.2 nmol/g ($n = 6$). Perfusion of isolated hearts with 1 mM l-homocysteine thiolactone and 10 μM adenosine caused tissue SAH levels to linearly increase. As shown in Fig. 5, the rate of increase was 3.8 nmol SAH/g/min. This is only about 7% of the maximal rate possible in the guinea pig heart, which was determined to be 59 nmol SAH/g/min (Deussen et al. 1987). Taking the V_{max} of the enzyme in vitro to be 2 nmol/min/mg protein (Schrader et al. 1981) and assuming a protein content of 29.5 mg/g heart, an identical V_{max} value can be calculated. Due to the similarity of the measured and calculated rates of SAH accumulation, it is very likely that the enzyme is not working under major constraints under in vivo conditions.

S-Adenosylhomocysteine as Index of Myocardial Ischemia

Whenever the oxygen supply to the heart via the coronary system becomes limited, oxidative phosphorylation is impaired and cardiac adenine nucleotides are dephosphorylated resulting in the formation of adenosine. Since tissue level of ATP is usually about 4000–6000 nmol/g and that of adenosine only 1–2 nmol/g, small changes in ATP may remain undetected but can cause large increases in the concentration of adenosine. Based on these considerations, studies carried out in dogs (Vrobel et al. 1982) and also in humans (Fox et al. 1974) revealed that release of adenosine into coronary venous blood was greatly augmented under conditions

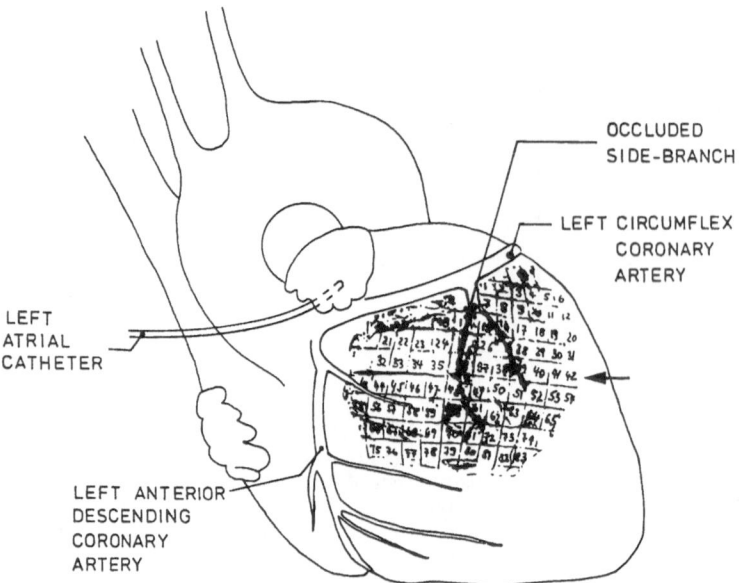

Fig. 6. Localization of subepicardial tissue samples analysed for adenosine and SAH in a representative dog experiment in which a side branch of the left circumflex coronary artery was occluded. Homocysteine (10 μmol/kg/min) was infused via the left atrial catheter. After 30 min of ischemia and homocysteine infusion the entire free wall of the left ventricle was rapidly removed, stop-frozen in liquid nitrogen, freezed-dried, and for mapping of metabolites subdivided into 83 subepicardial tissue samples. The *arrow* indicates the tissue slice, the analysis of which is shown in Fig. 7

of reduced oxygen supply to the heart. These studies are therefore consistant with the view that adenosine may be used as a sensitive index of cardiac hypoxia and ischemia. Given the relationship between adenosine, homocysteine, and SAH described above, it appears now feasable to trap part of the adenosine formed by the hypoxic heart by conversion into SAH. Due to the kinetics of SAH formation, this procedure amplifies the primary signal considerably (Deussen et al. 1987).

In order to evaluate when SAH can be used as a sensitive index of cardiac ischemia, experiments were carried out in the instrumented open chest of the dog (Fig. 6). Homocysteine thiolactone was continuously infused via the left arterial catheter into the left atrium at a rate of 10 μmol/kg/min. This elevated basal plasma level of homocysteine from normal 10 μM to the millimolar range, but did not alter basal coronary blood flow and contractile parameters of the heart. One minute after initiation of the homocysteine infusion, a small side branch of the left circumflex coronary artery was occluded. After 30 min of ischemia the entire free wall of the left ventricle was excised, frozen in liquid nitrogen, and freeze-dried. The excised tissue was then sectioned into 83 epicardial tissue samples (Fig. 6), which were analyzed for adenosine and SAH using sensitive HPLC techniques. Mapping of a representative tissue slice revealed (Fig. 7) that SAH was confined to the ischemic area. Within a small border zone, SAH increased from 1–3 nmol/g to over 100 nmol/g. The changes in global tissue adenosine, despite being the precur-

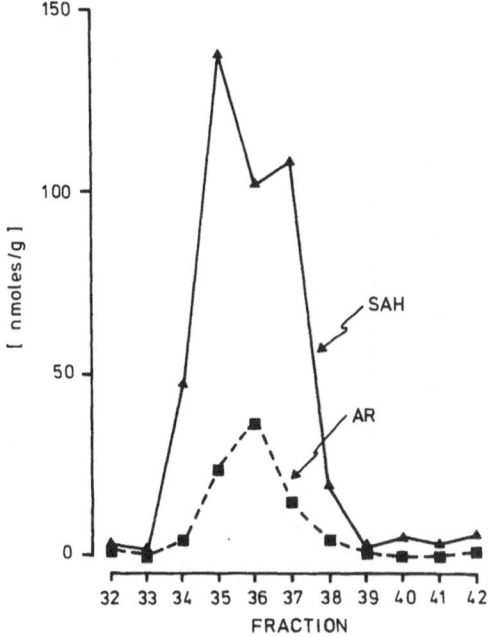

Fig. 7. Tissue levels of global adenosine and SAH in samples derived from the experiment described in Fig. 6. Sample number 36 was right beneath the occluded coronary artery. Note: In the ischemic border zone SAH levels are considerably increased, while in the same samples adenosine remains largely unaffected

sor of SAH, were less pronounced and compared with SAH adenosine less sharply defined the ischemic border zone.

The time course of SAH formation was studied in the same experimental model by taking multiple epicardial drill biopsy specimens in the center of the ischemic area as well as in the nonischemic control tissue. As is evident from data given in Fig. 8, SAH linearly accumulated in the ischemic tissue while in respective control samples SAH remained almost unchanged.

Using SAH as a sensitive marker of cardiac ischemia and hypoxia has certain experimental advantages over other istablished indices of cardiac ischemia, such as creatine phosphate and lactate. Basically, measurement of SAH reflects the kinetic process of conversion of adenosine into SAH, thereby leading to significant signal augmentation. Furthermore, SAH is a rather stabile metabolite, which is not critically affected by conventional tissue fixation procedures such as immersion in liquid nitrogen. This makes it possible to measure changes in SAH in different layers of the heart and to map epicardial and transmural changes of this purine compound.

To further explore the applicability of SAH, coronary perfusion pressure was reduced in the anesthetized open chest dog from normal 110 to 40 mmHg. Homocysteine was simultaneously administered and the transmural gradient of SAH in the left ventricular wall was determined. At 110 mmHg, SAH was found to be homogeneously distributed across the different layers of the left ventricular wall. At 40 mmHg, however, SAH steeply increased in the midportion of the ventricular wall reaching maximal values in the subendocardium (Fig. 9). From functional studies it is well established that at a coronary perfusion pressure of 40 mmHg

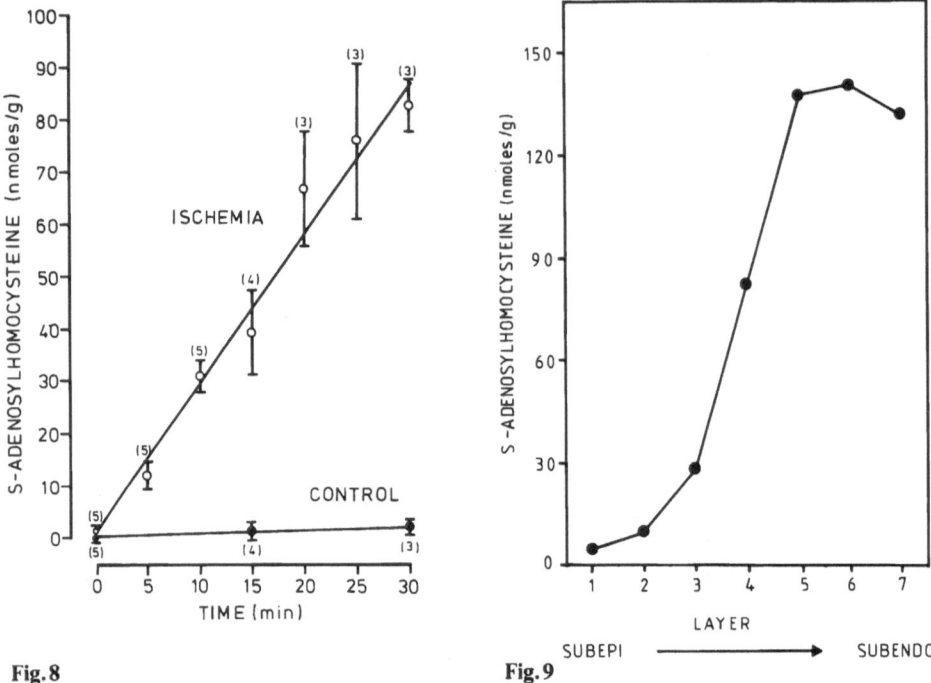

Fig. 8   **Fig. 9**

Fig. 8. Time course of SAH accumulation in severely ischemic and nonischemic (control) myocardium. Data are from five different experiments. Experimental conditions were as described in Fig. 6

Fig. 9. Transmural gradient of SAH in the left ventricular wall of the dog after reduction of coronary perfusion pressure to 40 mmHg for 30 min. Homocysteine was simultaneously infused at 10 μg/kg/min i. v. Thickness of the different myocardial layers analysed was 1.7–2.0 mm

coronary dilator reserve becomes exhausted and that coronary flow to the endocardium is first compromised (Bache and Dymek 1981). The transmural gradient of SAH provides the biochemical correlate for these previous findings.

Another interesting methodological aspect can be derived from our findings. Measurements of adenosine in epicardial transudates have been used in numerous studies as a measure of interstitial adenosine (Miller et al. 1979; Hanley et al. 1983; Kusachi and Olsson 1983). Our data, however, clearly demonstrate that epicardial SAH and thus free adenosine remain largely unchanged despite considerable changes in the deeper layers of the heart. Conversely, since SAH is a time-averaged signal, adenosine formed in the deeper layers of the heart does not reach the epicardial layers in significant amounts. Therefore epicardial transudate adenosine most likely reflect changes in the interstitial adenosine concentration of the superficial muscle layers only and is not representative of a mean interstitial adenosine of the entire ventricular wall.

Summary

Studies carried out in two experimental models, the buffer-perfused isolated guinea pig heart and the dog heart in situ, provide strong evidence that the accumulation of SAH in cardiac tissue is proportional to the free cytosolic concentration of adenosine and can be used as a sensitive index of coronary underperfusion. Several lines of evidence now suggest that any mismatch between myocardial oxygen supply and oxygen demand is first translated into changes of free adenosine, part of which is trapped as SAH when homocysteine is present. Formation of SAH ultimately reflects, via adenosine, changes in the energy status of the heart. Stated differently: SAH apparently can integrate all those cardiac events which finally lead to a reduction in oxidative phosphorylation either because of reduced tissue oxygenation or because of limited substrate supply. Due to the large gain in signal observed during ischemia and coronary underperfusion, SAH should be a suitable biochemical marker by which any improvement or deterioration of the oxygen supply/demand ratio and/or energy state of the heart can be monitored.

References

Bache RJ, Dymek DJ (1981) Local and regional regulation of coronary vascular tone. Cardiovascular diseases, vol XXIV (3). Grune and Stratton, New York, pp 191-212

Balaban RS, Kantor HL, Katz LA, Briggs RW (1986) Relation between work and phosphate metabolite in the in vivo paced mammalian heart. Science 232: 1121-1123

Bardenheuer H, Schrader J (1983) Relationship between myocardial oxygen consumption, coronary flow and adenosine release in the improved isolated working heart preparation of guinea pigs. Circ Res 51: 263-271

Bardenheuer H, Schrader J (1986) Supply-to-demand ratio for oxygen determines formation of adenosine by the heart. Am J Physiol 250: H173-H180

Becker BF, Gerlach E (1987) Nachweis multipler hypoxischer Areale in "normoxisch" perfundierten isolierten Herzen. Z Kardiol 76: 254

Belloni FL, Bruttig SP, Rubio R, Berne RM (1986) Uptake and release of adenosine by cultured rat aortic smooth muscle. Microvasc Res 32: 200-210

Berne RM (1963) Cardiac nucleotides in hypoxia: possible role in regulation of coronary blood flow. Am J Physiol 204: 317-322

Berne RM (1980) The role of adenosine in the regulation of coronary blood flow. Circ Res 47: 807-813

Berne RM, Rubio R (1979) Coronary circulation. In: Berne RM, Sperelakis N (eds) The cardiovascular system, sect 2, vol I. American Physiological Society, Washington DC, pp 873-952 (Handbook of physiology)

Bünger R, Soboll S (1986) Cytosolic adenylates and adenosine release in perfused working heart. Eur J Biochem 159: 203-213

De la Haba G, Cantoni GL (1959) The enzymatic synthesis of S-adenosylhomocysteine from adenosine and homocysteine. J Biol Chem 234: 603-608

Deussen A, Möser G, Schrader J (1986) Contribution of coronary endothelial cells to cardiac adenosine production. Pflugers Arch 406: 608-614

Deussen A, Borst M, Schrader J (1988) Formation of S-adenosylhomocysteine in the heart. I. An index of free intracellular adenosine. Circ Res (in press)

Drury AN, Szent-Györgyi A (1929) The physiological activity of adenine compounds with especial reference to their action upon the mammalian heart. 7 Physiol (London) 68: 213-237

Feigl EO (1983) Coronary physiology. Physiol Rev 63: 1-205

Forrester T, Williams CA (1977) Release of adenosine triphosphate from isolated adult heart cells in response to hypoxia. J Physiol (Lond) 268: 371-390

Fox AC, Reed GE, Glassman E, Kaltman AJ, Silk BB (1974) Release of adenosine from human hearts during angina induced by rapid atrial pacing. J Clin Invest 53: 1447-1457

Fredholm BB, Hedquvist P, Lindstrom K, Wennmalm M (1982) Release of nucleosides and nucleotides from the rabbit heart by sympathetik nerve stimulation. Acta Physiol Scand 116: 285-295

Fukami MH, Holmsen H, Salganicoff L (1976) Adenine nucleotide metabolism of blood platelets. IX. Time course of secretion and changes in energy metabolism in thrombin-treated platelets. Biochim Biophys Acta 444: 633-643

Gerlach E, Deuticke B, Dreisbach RH (1963) Der Nucleotid-Abbau im Herzmuskel bei Sauerstoffmangel und seine mögliche Bedeutung für die Coronardurchblutung. Naturwissenschaften 50: 228-229

Gordon EL, Pearson JD, Slakey LL (1986) The hydrolysis of extracellular adenine nucleotides by cultured endothelial cells from pig aorta. J Biol Chem 261: 15496-15504

Hanley F, Messina LM, Baer RW, Uhlig PN, Hoffman JIE (1983) Direct measurement of left ventricular interstitial adenosine. Am 7 Physiol 245: H 327-335

Jacobus WE (1985) Respiratory control and the integration of heart high energy metabolism by mitochondrial creatine kinase. Annu Rev Physiol 47: 707-725

Kroll K, Schrader J, Piper HM, Henrich M (1987) Release of adenosine and cyclic AMP from coronary endothelium in isolated guinea pig hearts: relation to coronary flow. Circ Res 60: pp659-665

Kusachi S, Olsson RA (1983) Pericardial superperfusion to measure cardiac interstitial adenosine concentration. Am J Physiol 244: H458-H461

Lloyd HGE, Schrader J (1987) The importance of the transmethylation pathway for adenosine metabolism in the heart. In: Gerlach E, Becker BF (eds) Topics and perspectives in adenosine research. Springer, Berlin Heidelberg New York, pp 199-207

Manfredi JP, Sparks HV (1982) Adenosine's role in coronary vasodilation induced by atrial pacing and norepinephrine. Am J Physiol 243: H536-H543

McCormack JG, Denton RM (1986) Ca^{++} as a second messenger within mitochandria. TIBS 11: 258-262

Miller WL, Belardinelli L, Bacchus A, Foley DH, Rubio R, Berne RM (1979) Canine myocardial adenosine and lactate production, oxygen consumption, and coronary blood flow during stellate ganglia stimulation. Circ Res 45: 24-29

Mistry G, Drummond GJ (1986) Adenosine metabolism in microvessels from heart and brain. J Mol Cell Cardiol 18: 13-22

Newby AC, Holmquist CA, Illingworth JA, et al. (1983) The control of adenosine concentration in polymorphonuclear leukocytes, cultured heart cells and isolated perfused heart from the rat. Biochem J 214: 317-323

Nuutinen EM, Wilson DF, Erecinska M (1984) Mitochondrial oxidative phosphorylation: tissue oxygen sensor for regulation of coronary flow. Adv Exp Med Biol 169: 351-357

Olsson RA, Bünger R (1987) Metabolic control of coronary blood flow. Prog Cardiovasc Res 29: 369-387

Olsson RA, Saito D, Steinhart CR (1982) Compartmentalisation of the adenosine pool of dog and rat hearts. Circ Res 50: 617-626

Pearson JD, Carleton JS, Gordon JL (1980) Metabolism of adenine nucleotides by ectoenzymes of vascular endothelial and smooth muscle cells in culture. Biochem J 190: 421-429

Schrader J (1983) Metabolism of adenosine and sites of production in the heart. In: Berne RM, Rall TW, Rubio R (eds) Regulatory function of adenosine. Martinus Nijhoff The Hague, pp 133-156

Schrader J, Schütz W, Bardenheuer H (1981) Role of S-adenosylhomocysteine hydrolase in adenosine metabolism in mammalian heart. Biochem J 196: 65-70

Schütz W, Schrader J, Gerlach E (1981) Different sites of adenosine formation in the heart. Am J Physiol 240: H963-H970

Sparks HV, Bardenheuer H (1986) Regulation of adenosine formation by the heart. Circ Res 58: 193-201

Ueland PM (1982) Pharmacological and biochemical aspects of S-adenosylhomocysteine and S-adenosylhomocysteine hydrolase. Pharmacol Rev 34: 223-253

Vrobel ThR, Jorgensen CR, Bache RJ (1982) Myocardial lactate and adenosine metabolite production as indicators of exercise-induced myocardial ischemia in the dog. Circulation 66: 555-564

IV. Nervous Systems

Changes of the Bioelectrical Activity and Extracellular Micromilieu in the Central Nervous System During Variations of Local Oxygen Pressure

E.-J. Speckmann, D. Bingmann[1], A. Lehmenkühler, and H. G. Lipinski[2]

Institut für Physiologie, Universität Münster, Robert-Koch-Str. 27a, 4400 Münster, FRG

The present chapter gives a brief overview on our experimental observations concerning the effects of changes in oxygen pressure on neuronal functioning. In the experiments mentioned the oxygen pressure in nervous tissue was shifted to hypoxic and hyperoxic levels by various methods and techniques. In a variety of the described situations shifts of oxygen pressure caused simultaneous changes of carbon dioxide pressure and of H^+ concentration which are due, e.g., to disturbances of microcirculation and/or to titration of bicarbonate. Therefore, a paragraph dealing with the actions of hypercapnia and acidosis is intercalated between the paragraphs devoted to effects of hypoxia and hyperoxia.

The experiments were performed in the motor cortex and the spinal cord of anesthetized and artificially ventilated rats and cats, in hippocampal slices of guinea pigs, and in sensory spinal ganglion cells of rats grown in tissue culture [for methods see 4, 9, 20, 22, 25, 29, 30].

Decrease in Oxygen Pressure (Hypoxia)

It is well known that the oxygen pressure in central nervous tissue decreases when the oxygen content of the artificially ventilated gas mixture is lowered in the anesthetized animal. However, it should be noted in this context that the absolute decrease in oxygen pressure varies in a wide range, dependent on the location of the pO_2 sensor in the tissue (for oxygen pressure fields in the central nervous system cf. [26]).

A lowering of the oxygen pressure in tissue leads to a progressive depolarization of neocortical and spinal neurons provided that the carbon dioxide pressure is kept constant (Fig. 1 A). The membrane resistance measured by current pulses injected through the microelectrode is usually found to decrease in such conditions of isolated hypoxia. Especially, in interneurons the decrease of membrane

[1] Present address: Poliklinik für zahnärztliche Chirurgie der Universität, Augustusplatz 2, 6500 Mainz, FRG

[2] Present address: Neurologische Klinik der Technischen Universität, Möhlstraße 28, 8000 München, FRG

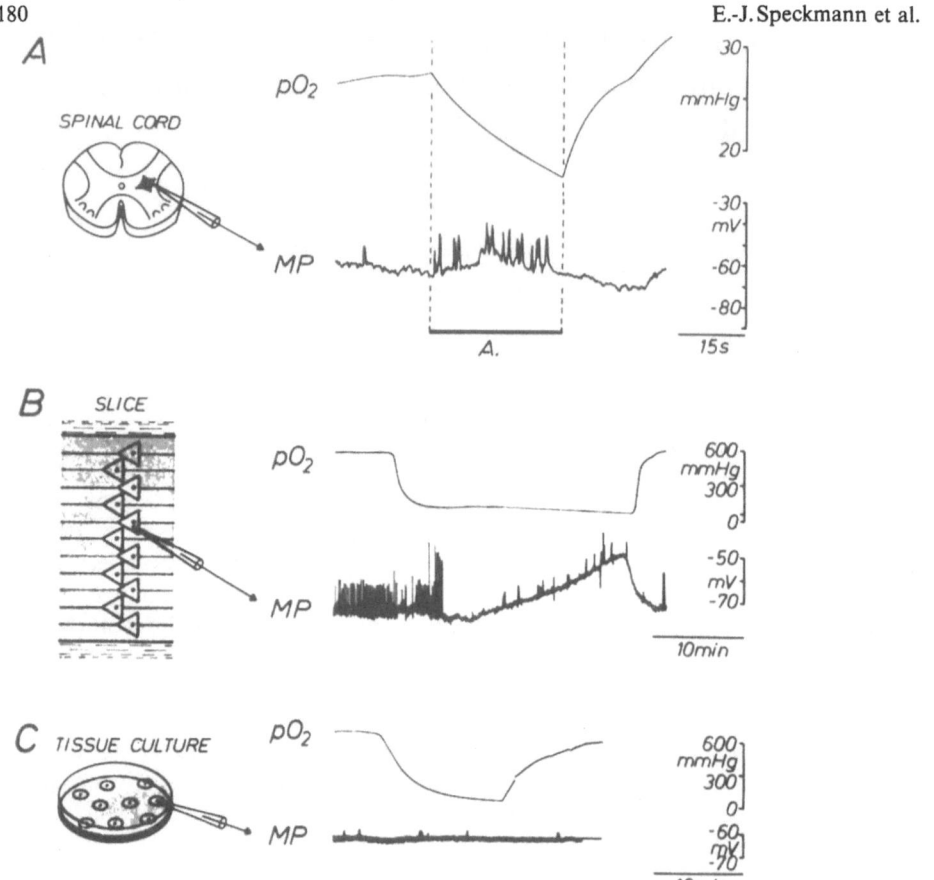

Fig. 1 A–C. Different effects of lowering PO_2 on membrane potentials *(MP)* of neurons in vivo (**A**) and in vitro (**B, C**). **A** Changes of MP of a spinal cord interneuron (rat) and of tissue PO_2 measured in the vicinity of the neuron during a 30-s ventilatory arrest (**A**). **B** MP changes of a CA3 neuron in a 400-μm thick hippocampal slice (guinea pig) during shifts of the pO_2 in the bath. **C** Missing MP changes of a sensory spinal ganglion cell (rat) grown in tissue culture during shifts of the PO_2 in the superfusate. (**A** from [30]; **B** from [9]; **C** from [1]

potential is superimposed with spontaneously occurring and long-lasting excitatory postsynaptic potentials which are enhanced in amplitude and frequency (Fig. 1 A). These neuronal reactions are associated initially with an increase of discharge frequency. With progressing hypoxia action potentials regularly show a partial blockade before spike generation fails. In this stage excitatory postsynaptic potentials can still be evoked or may occur spontaneously. As a whole, increasing hypoxia induces progressive depolarization of neocortical and spinal neurons associated with biphasic changes in discharge frequency [1, 19, 29, 30, 31, 32].

For a more detailed analysis of the dependency of neuronal activity on local oxygen pressure, experimental conditions more controlled appear to be a prerequisite. Such conditions can be realized partly in hippocampal slice preparations because (a) the excised tissue is superfused in vitro with a defined salt solution with constant ion concentrations; (b) problems arising from changes of microcirculation are eliminated; and (c) pO_2-dependent fluctuations of the synaptic input probably are of minor importance.

Neurons impaled in the CA3 region of hippocampal slices show a continuous depolarization which starts about 10 min after the commencement of the decrease in the bath oxygen pressure (Fig. 1 B). The depolarization is often preceded by a small transient hyperpolarization. The rate of depolarization is clearly related to the recording depth. It is slowed down in cells impaled near the surface of the slice [3, 4, 9].

The described sensitivity of CA3 neurons in hippocampal slices to changes of oxygen pressure contrasts strikingly the poor dependency of membrane potential of sensory spinal ganglion cells grown in tissue culture on variations of oxygen pressure in vitro. Figure 1 C demonstrates that shifts of the oxygen pressure in the bath fluid which are found to clearly modify the activity of CA3 neurons in slice preparations hardly affected the resting membrane potential of cultured sensory neurons. Furthermore, significant alterations of the membrane resistance or of the amplitude, duration, steepness, and other parameters of action potentials are missing [1, 3].

As already described, neocortical and spinal neurons depolarize after a reduction of the local oxygen pressure provided that the lowering of oxygen pressure in tissue remained free of essential secondary changes in carbon dioxide pressure and H^+ concentration. The responses of the neuronal membrane potential in vivo are found to be associated with an increase in the extracellular K^+ concentration and a negative-going shift of the cortical DC potential (Fig. 2 A; cf. 7 A). In the hippocampal slice variations of the oxygen pressure also induce a rise in the extracellular K^+ concentration which is dependent on the amount of decrease in oxygen pressure and on the depth of recording below surface (Fig. 2 B). Thus, the increase in extracellular K^+ concentration is related to the concomitant decrease in neuronal membrane potential. In this context it should be mentioned that the extracellular milieu of sensory spinal ganglion cells in tissue culture is "clamped" by superfusing the neurons directly [24, 25, 30].

The findings presented demonstrate that the extracellular concentration of K^+ increases in the tissue of the central nervous system when the local oxygen pressure is lowered. This variation in the microenvironment is reflected in reactions of neocortical glial cells, the membrane potential of which is determined predominantly by the extracellular K^+ concentration. A typical experiment is displayed in Fig. 3. It shows that with a decrease of oxygen pressure in tissue during a short ventilatory arrest, a depolarization of the glial cell takes place [17].

In summary, the findings suggest that neither the oxygen deficiency nor a critically low oxygen pressure per se are exclusively responsible for the observed effects on neurons and glial cells in tissue. Other mechanisms seem also to contribute to these effects. Among these, changes of the extracellular microenvironment, especially the increase in the extracellular K^+ concentration due to an increase in unspecific ion permeabilities [27] and to a slowing down of ion pumps, may play an important role for the hypoxic depolarization of neurons in the central nervous system.

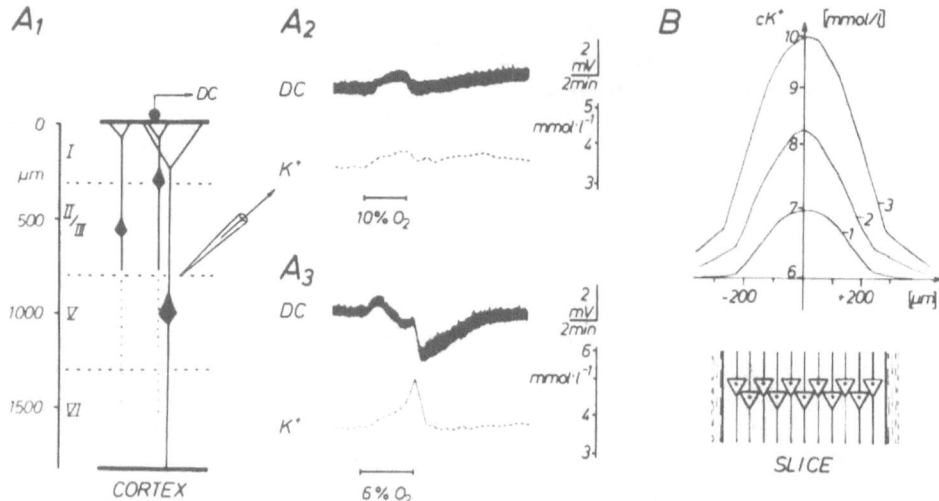

Fig. 2 A, B. Elevation of extracellular potassium concentration in the neocortex (rat) in vivo (**A**) and in a hippocampal slice (guinea pig) in vitro (**B**) with lowering pO_2. **A₁** Positions of the electrodes recording DC potentials *(DC)* and the extracellular potassium concentration (K⁺). **A₂,₃** Epicortical DC potential and intracortical K⁺ concentration during lowering of the inspiratory O_2 concentration from 20% to 10% (**A₂**) and to 6% (**A₃**). **B** Depth profiles of the extracellular K⁺ concentrations in a 500-μm thick hippocampal slice. Potassium profiles were recorded at bath pO_2 values of 600 mmHg *(1)*, 150 mmHg *(2)*, and 70 mmHg *(3)*. (**A** from [24]; **B** from [25])

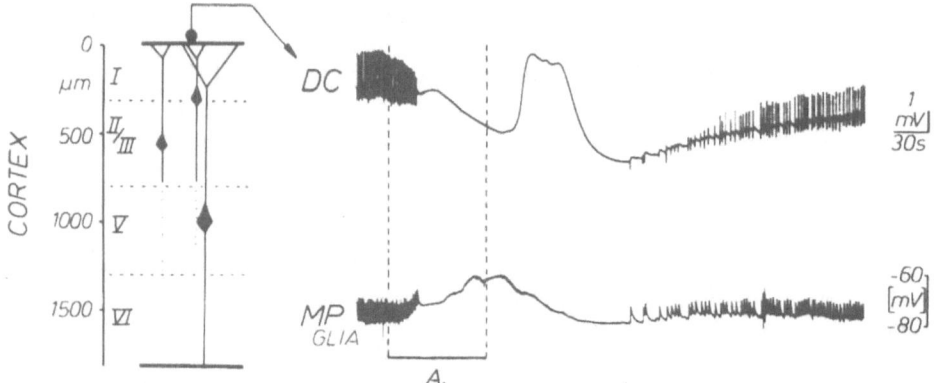

Fig. 3. Changes of the epicortical DC potential *(DC)* and of the membrane potential *(MP)* of a glial cell impaled at a depth of 1000 μm below the cortical surface during a 100-s ventilatory arrest *(A)*. Anesthetized and artificially ventilated rat

Increase in Carbon Dioxide Pressure and in H⁺ Concentration (Hypercapnia and Acidosis)

With a progression of oxygen deficiency the negative shift of the DC potential at the cortical surface, which represents a hypoxic response (see above; cf. also Fig. 7 A) turns to a positive deflection (Fig. 4 A). Such diphasic effects of hypoxia

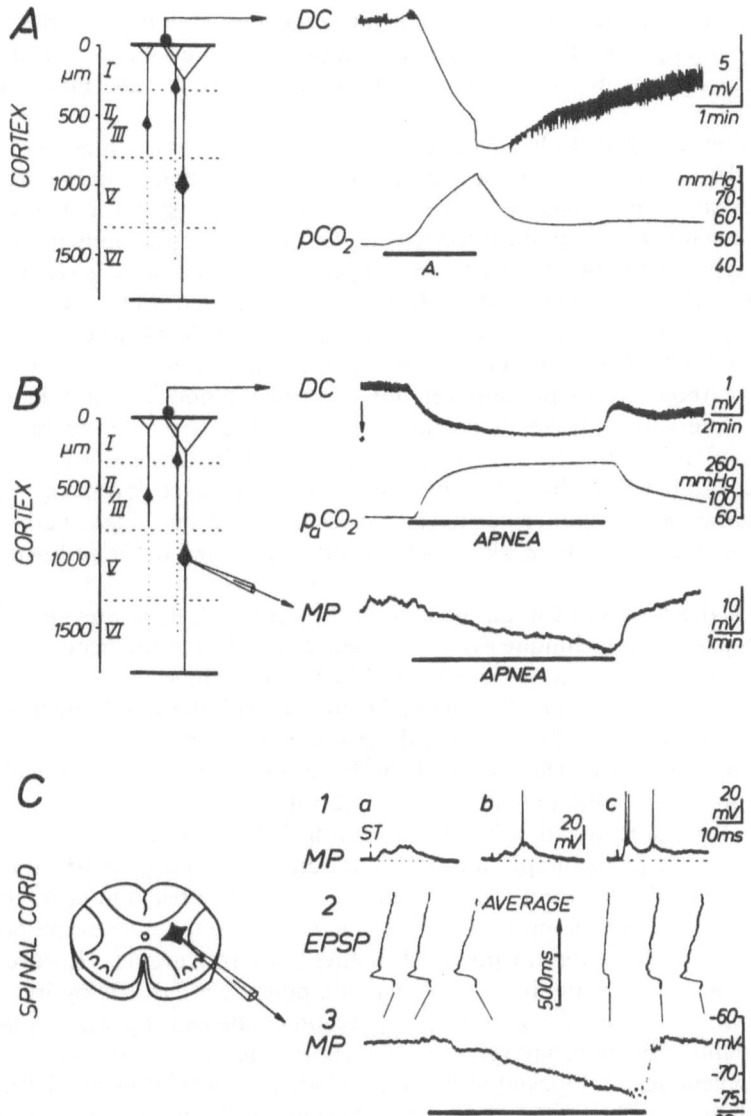

Fig. 4 A–C. Changes of epicortical DC potentials (**A, B**; *DC*), of pCO₂ in cortical tissue (**A**) and in the carotid artery (**B**) and of membrane potential *(MP)* of neurons impaled in the cortex (**B**) and in the spinal cord (**C**) during ventilatory arrests after breathing room air (**A**: *A*) or pure oxygen (**B, C**: *apnea*). Anesthetized and artificially ventilated rat (**A, C**) and cat (**C**). **C1** Sequences of excitatory synaptic potentials *(EPSP)* elicited by stimulation (ST; increasing intensity: *a–c*) of the sensorimotor cortex under control conditions. **C2** Changes of averaged (*n* = 20) EPSP following cortical stimulation during apnea. (**C** from [10])

are always accompanied by a lowering of heart frequency and fall of arterial blood pressure. Among other things the latter reactions lead to an increase of the local carbon dioxide pressure and, thus, to a positive shift of the DC potential (Fig. 4 A; [10, 30]).

To study the effect of an isolated increase in the carbon dioxide pressure in nervous tissue, the following technique labeled apnea was applied to induce hypercapnia. After the animal has been ventilated with pure oxygen for at least 30 min, artificial ventilation is interrupted leaving the trachea of the animal connected to the oxygen container. Under these conditions oxygen uptake is secured by convection into the lungs while carbon dioxide is not removed. In this way the arterial oxygen pressure is declining very slowly from a high level, whereas the carbon dioxide pressure rises continuously to considerable values (Fig. 4 B). Due to the increased carbon dioxide pressure cerebral and spinal blood flow and therefore the oxygen pressure in these tissues are enhanced. In a later period of an apnea (15–20 min) the initial rise of the oxygen pressure in tissue is followed by a slight reduction. Using this procedure high carbon dioxide levels can be reached in tissue avoiding subnormal oxygen pressures. The described method was used in the experiments the results of which will be reported in the following [28].

In the anesthetized rat and cat, an increase in the carbon dioxide pressure by the aforementioned apnea technique evokes a positive DC shift at the neocortical surface. An example of this finding is displayed in Fig. 4 B. The tracings demonstrate that the displacement of the DC potential coincides with the well-known depression of the conventional EEG. During these events neocortical and most of spinal neurons hyperpolarize. The extent of the hyperpolarization is reciprocally related to the initial membrane potential [10, 11, 12, 14].

The neuronal hyperpolarization occurring during hypercapnia is associated with a reduction of steepness, amplitude, and duration of excitatory postsynaptic potentials. A sample of typical tracings is given in Fig. 4 C. The reaction of inhibitory postsynaptic potentials is more difficult to analyze since the membrane potential is shifted to the equilibrium potential of these synaptic events. However, comparing the alterations of inhibitory postsynaptic potentials induced by intracellular current injection with the actions of hypercapnia, one can suggest that inhibitory postsynaptic potentials are reduced, as well. Furthermore, antidromically evoked action potentials are blocked at carbon dioxide pressure levels exceeding 100 mm Hg. Simultaneously, the repetition rate of neuronal discharges elicited by depolarizing current injection is reduced. As a whole, these findings show that carbon dioxide depresses neuronal discharges both by direct actions on the cell membrane and via synaptic processes [10, 31, 32].

In the hippocampal slice preparation, an elevation of carbon dioxide content in the bath fluid is able to induce neuronal depolarizations as well as hyperpolarizations (Fig. 5 A). The direction of the response to hypercapnia depends on the location of the neurons impaled. Whereas neurons near the boundary zone of tissue and superfusate depolarize, neurons in the innermost layers of the slice hyperpolarize (Fig. 5 B). Independent of the direction of the carbon dioxide-induced change in membrane potential the membrane resistance was most often reduced [2, 16].

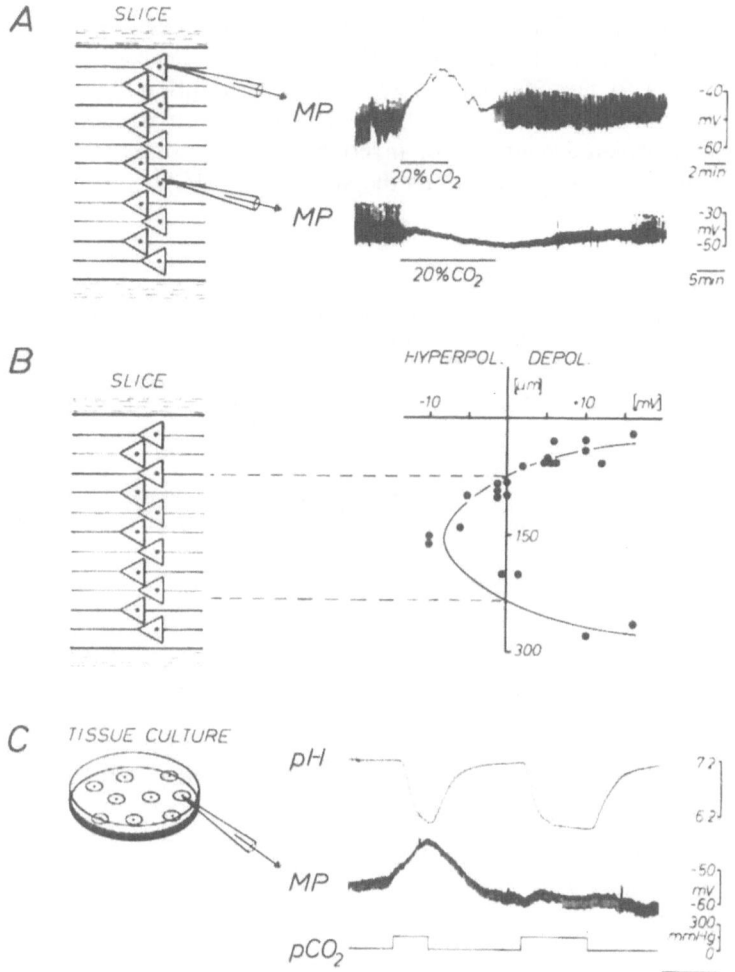

Fig. 5 A–C. Changes of the membrane potentials *(MP)* of neurons in vitro during elevation of pCO_2 in the bath. **A** Depolarization of a superficial neuron of a hippocampal slice (near the bath) and hyperpolarization of a neuron in the central zone of the slice (remote from the bath) during superfusion of the slice with a salt solution equilibrated with 20% CO_2 instead of 5%; guinea pig. **B** Dependency of the extent and of the direction of the MP shift on the depth of the neuron within the slice during exposition to a bath being equilibrated with 20% CO_2 instead of 5% CO_2. **C** pCO_2/pH induced depolarizations of an isolated sensory spinal ganglion cell (rat, tissue culture) exposed directly to the superfusate. (**B** from [16])

In cultured sensory spinal ganglion cells, an acidification induced by adding HCl to the bath fluid leads to a decrease of the membrane potential. Similar effects were observed during a primary rise of the carbon dioxide pressure (Fig. 5 C; [5, 24]).

The different neuronal responses to hypercapnia and/or acidosis suggest that different buffer capacities at the sites of the tested neurons modify the actual reaction essentially. The resulting higher acidification in the outer layers as compared with the inner ones of a slice may lead to an H^+-dependent unspecific cation inward current and, thus, to neuronal depolarization. The assumption that the actual buffer capacity contributes markedly to the neuronal reaction is supported by findings obtained in isolated neurones of invertebrates and in cultured sensory spinal ganglion cells. In these preparations the addition of HCO^-_3 or of macro-

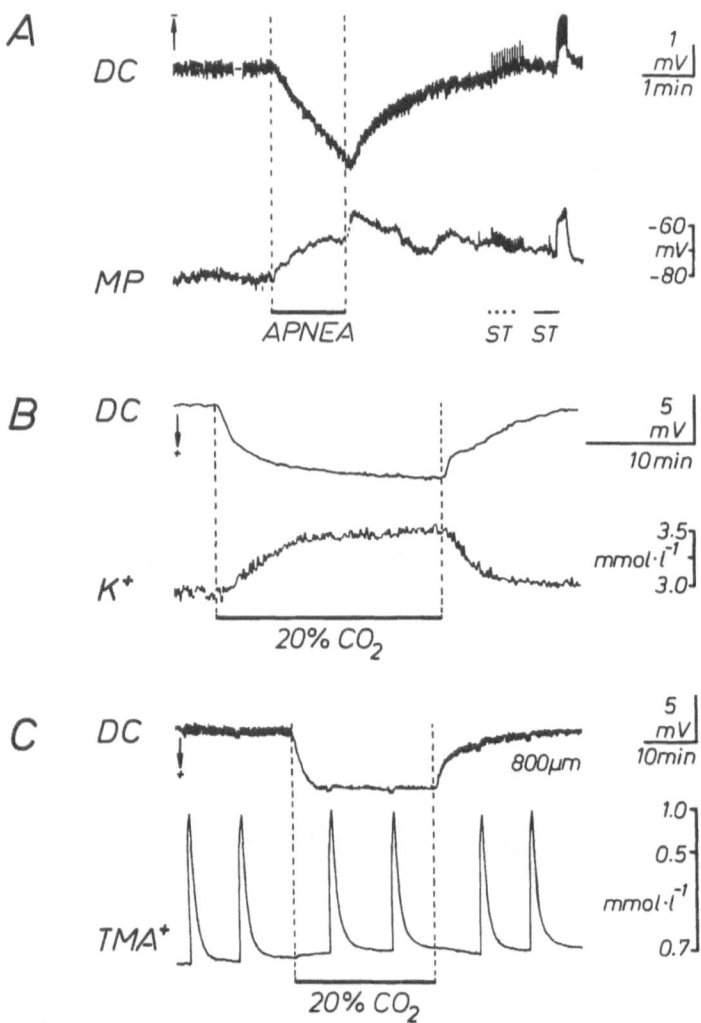

Fig.6A-C. Effects of hypercapnia on DC potentials (A-C), on the membrane potential of a cortical glia cell (A), on the extracellular potassium concentration (K^+; B) and on the extracellular volume fraction monitored by repeated applications of tetramethylammonium (TMA^+; C); anesthetized and artificially ventilated rat. Hypercapnia was achieved in A by a ventilatory arrest after breathing pure oxygen *(APNEA)*, in B, C by an elevation of the inspiratory CO_2 content to 20%. *ST;* low frequency *(dotted line)* and high frequency *(solid line)* electrical stimulation of the cortical surface. (A from [17]; B, C from [23])

molecular buffer substances like hemoglobin reverses the primary depolarizations to hyperpolarizations. In this context the observation should be taken into account that the depolarization of sensory spinal ganglion cells occurring during a second hypercapnic period is small as compared with the neuronal reaction during the first one (Fig. 5C). This can be attributed to an elevation of the intracellular buffer capacity outlasting the first rise in carbon dioxide pressure [2, 16, 24, 33].

Glial cells in the neocortex depolarize during hypercapnia (Fig. 6A). This response was observed in all neocortical layers with differences in amplitude being missing. The reaction of glial cells to hypercapnia can be interpreted as being due to the elevation of the extracellular K^+ concentration occurring with the increase in the carbon dioxide pressure (Fig. 6B). The size of the extracellular volume fraction is found to be unaffected during the hypercapnic period. Since the K^+ permeability of the blood-brain barrier is lowered during hypercapnia it can be assumed that K^+ is released from brain cells [15, 17, 18; see also 6, 7].

In summary, the observations described suggest that the carbon dioxide-induced hyperpolarization of neurons is caused by an increase of the K^+ conductance. This effect is probably the consequence of an H^+-dependent rise in the intracellular Ca^{2+} activity.

Increase in Oxygen Pressure (Hyperoxia)

Elevations of the inspiratory oxygen content above normal levels elicit distinct deviations of the neocortical DC potential in the anesthetized rat and cat. In animals ventilated artificially at a constant rate, the first response consists, as a rule, in a surface-positive DC shift which increases in amplitude with a further rise of the oxygen content in the inspiratory gas mixture. The primary positive deflection may turn to a slowly developing negative DC deviation. A sample of recordings is displayed in Fig. 7 [11, 13, 29].

The initial positive DC shift cannot be attributed to direct actions of oxygen on the generator structures of the DC potential, but seems to depend on an increase of the carbon dioxide pressure resulting predominantly from higher concentrations of oxygenated hemoglobin in the venous blood. This interpretation is supported by a number of experimental results. Thus, continuous measurements in the carotid arteries and in the jugular veins confirm that each enhancement of the inspiratory oxygen content is accompanied by an increase of the carbon dioxide pressure in the blood provided that the rate of ventilation is kept constant. In accordance with this observation, cortical blood flow is usually found to rise [11, 13, 29].

The primary positive deviation of the cortical DC potential proceeds to a slowly rising negative shift already at oxygen pressure ranging below 700 mm Hg. Comparative measurements show that the critical oxygen pressure in the inspiratory gas mixture at which such a transition of the DC response takes place may vary within a wide range. The decisive factor seems to consist in the actual rise of the oxygen pressure in tissue which is determined by a number of additional mechanisms such as general cortical blood flow, changes in local microcirculation, and in oxygen consumption. This conclusion can be drawn from continuous re-

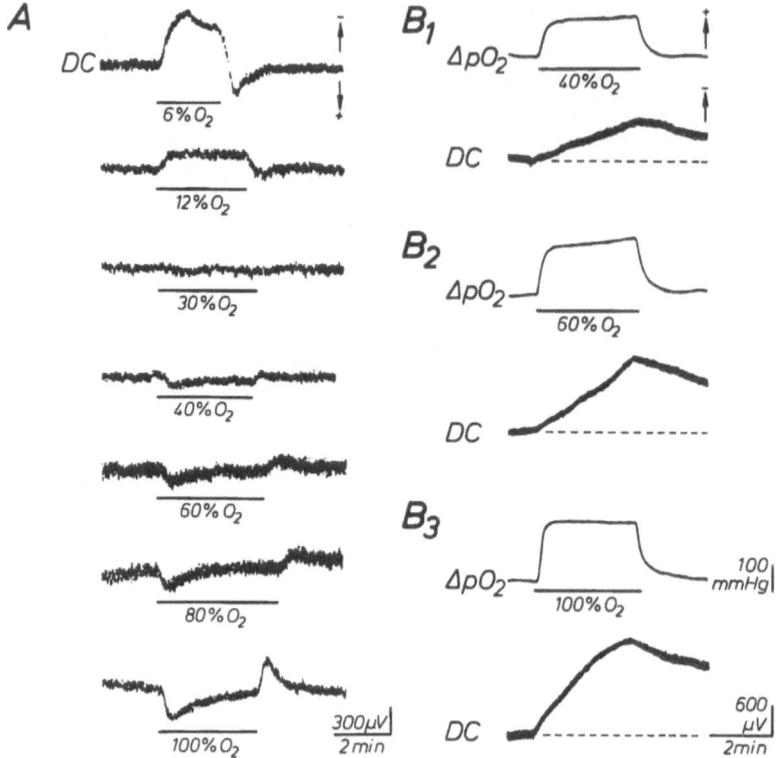

Fig. 7 A, B. Changes of the epicortical DC potential *(DC)* after switching the inspiratory O_2 content from normoxic to hypo-(**A**) and hyperoxic (**A, B$_{1-3}$**) values; anesthetized and artificially ventilated rat. **B$_{1-3}$** Simultaneous recordings of DC shifts and of tissue pO_2 after a rise of the inspiratory O_2 content to 40% (**B$_1$**), 60% (**B$_2$**) and 100% (**B$_3$**). [11]

cordings of the DC potential and of the oxygen pressure in cortical tissues on various experimental conditions. As shown in Fig. 7B the steepness of the negative shift of the DC potential is clearly related to the actual elevation of the oxygen pressure level at the cortical surface [11, 13, 29].

The effects of elevated oxygen pressures on the bioelectrical activity of central nervous structures become more evident if hyperbaric oxygenation is applied. To this purpose animals are ventilated in a chamber allowing treatment with oxygen high pressure. A typical experiment is displayed in Fig. 8A. The compression of ambient oxygen to 9 atmospheres elicits an initial increase of oxygen pressure in tissue and a secondary decline. In this case, a further rise of the oxygen pressure can be seen about 10 min later. This increase in oxygen pressure parallels the onset of seizure discharges [21].

Investigations performed in hippocampal slice preparations provide further information on the mechanisms underlying the hyperoxic effects described. With a rise of the oxygen pressure in the bath fluid the majority of neurons depolarize and the membrane resistance is elevated. Typical examples of such long-lasting depolarizations are given in Fig. 8B. The shifts in membrane potential and mem-

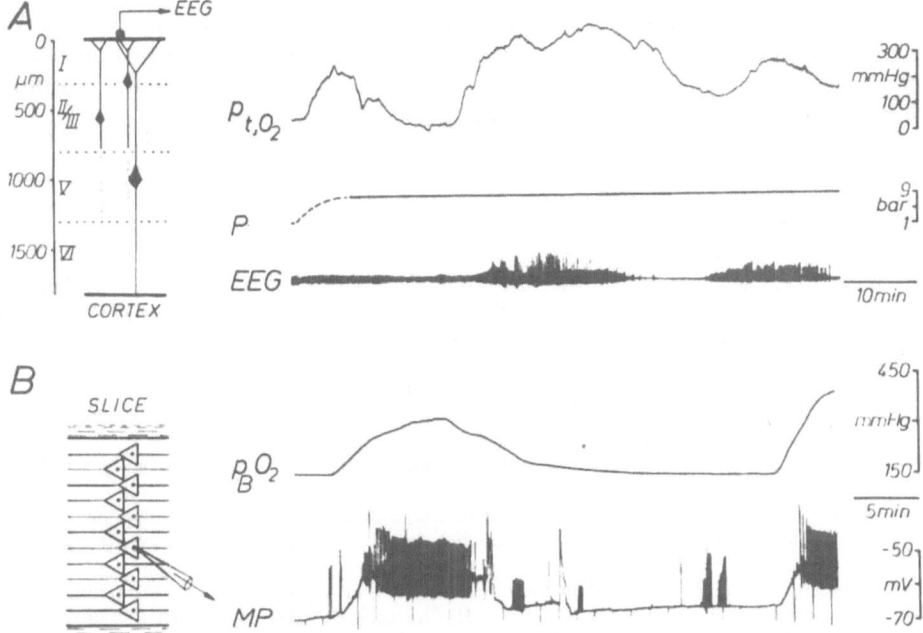

Fig. 8 A, B. Seizure discharges in the epicortical EEG and in the recording of the membrane potential of a CA3 neuron of a hippocampal slice during exposure to abnormally high pO_2 values; rat. **A** Elevated tissue pO_2 values (p_t, O_2) were achieved by exposing the rat being artificially ventilated with pure oxygen to 9 bar ambient pressure *(P)*. **B** Elevated pO_2 values in the bath ($p_B O_2$) were achieved by superfusing slices with a Hepes-buffered salt solution equilibrated with 100% O_2. Hyperoxia led to depolarizations and to a rise of the membrane resistance, which was monitored by repeated intracellular injections of hyperpolarizing currents of 0.2 nA *(downward strokes)*. (**A** from [21]; **B** from [8])

brane resistance are found to be widely constant in single cells when the effects of repetitive elevations in oxygen pressure are tested. In different cells, however, the fluctuations of membrane potential and membrane resistance vary considerably. In any case, the shifts of membrane potential are clearly related to the changes in membrane resistance. The increase in membrane resistance may be interpreted to represent a decrease in K^+ conductance [8].

In summary, it is suggested that the increase in membrane resistance is the main cause for the depolarization of neurons during hyperoxia. This depolarization may be assumed to be reflected in the negative displacement of the cortical DC potential and to be the basis of the initiation of seizure discharges.

Conclusions

Independent of a further clarification of the mechanisms involved in changes of neuronal activity during variations of gas pressures in central nervous tissues, it can be assumed that (a) an increase in the extracellular K^+ concentration plays an important role in neuronal depolarization during hypoxia; (b) a selective increase in K^+ conductance contributes essentially to neuronal hyperpolarization during hypercapnia and acidosis; and (c) a decrease in membrane permeability is at least initially responsible for neuronal depolarization during hyperoxia.

References

1 Bingmann D, Kienecker EW (1984) Effects of hypoxia on regenerated sinus nerve fibres in vivo and on neurons in vitro. In: Pallot D (ed) The peripheral arterial chemoreceptors. Croom Helm, London, pp 243-252

2 Bingmann D, Kolde G (1981a) Reactions of CA3 neurons in hippocampal slices to changes of CO_2 and pH in the bath solution. Pflugers Arch S391: R32

3 Bingmann D, Kolde G (1981b) Reactions of neurons in vitro to changes of PO_2 in the bath solution. Pflugers Arch S391: R32

4 Bingmann D, Kolde G (1982) PO_2-profiles in hippocampal slices of the guinea pig. Exp Brain Res 48: 89-96

5 Bingmann D, Pietruschka F (1978) Effects of CO_2 and H^+ ions on the resting membrane potential and on action potentials of mammalian dorsal root ganglion cells grown in tissue culture. Pflugers Arch S377: R43

6 Bingmann D, Kienecker EW, Knoche H (1977) Chemoreceptor activity in the rabbit carotid sinus nerve during regeneration. In: Acker H, Fidone S, Pallot D, Eyzaguirre C, Lübbers DW, Torrance RW (eds) Chemoreception of the carotid body. Springer, Berlin Heidelberg New York, pp 36-43

7 Bingmann D, Kienecker EW, Caspers H, Knoche H (1981) Chemoreceptor activity of sinus nerve fibres after their implantation into the wall of the external carotid artery. In: Belmonte C, Pallot D, Acker H, Fidone S (eds) Arterial chemoreceptors. Leicester University Press, pp 92-101

8 Bingmann D, Kolde G, Speckmann EJ (1982) Effects of elevated PO_2-values in the superfusate on neuronal activity in hippocampal slices. In: Klee MR, Lux HD, Speckmann EJ (eds) Physiology and pharmacology of epileptogenic phenomena. Raven, New York, pp 97-104

9 Bingmann D, Kolde G, Lipinski HG (1984) Relations between PO_2 and neuronal activity in hippocampal slices. In: Lübbers DW, Acker H, Leniger-Follert E, Goldstick TK (eds) Oxygen transport to tissue V. Plenum, New York, pp 215-226

10 Caspers H, Speckmann EJ (1972) Cerebral pO_2, pCO_2 and pH: Changes during convulsive activity and their significance for spontaneous arrest of seizures. Epilepsia 13: 699-725

11 Caspers H, Speckmann EJ (1974) Cortical DC shifts associated with changes of gas tensions in blood and tissue. In: Remond A (ed) Handbook of electroencephalography and clinical neurophysiology, vol 10/A. Elsevier, Amsterdam, pp 41-65

12 Caspers H, Schütz E, Speckmann EJ (1963) Gleichspannungsveränderungen an der Hirnrinde bei Sauerstoffmangel. Z Biol (Munich) 114: 112-126

13 Caspers H, Speckmann EJ, Simmich W, Zoll WR (1974) Beeinflussung der Sauerstoffintoxikation des Zentralnervensystems durch Carboanhydrasehemmstoffe. Res Exp Med (Berl) 163: 125-136

14 Caspers H, Speckmann EJ, Lehmenkühler A (1979) Effects of CO_2 on cortical field potentials in relation to neuronal activity. In: Speckmann EJ, Caspers H (eds) Origin of cerebral field potentials. Thieme, Stuttgart, pp 151-163

15 Caspers H, Speckmann EJ, Lehmenkühler A (1980) Electrogenesis of cortical DC potentials. In: Kornhuber HH, Deecke L (eds) Motivation, motor and sensory processes of the brain: electrical potentials, behaviour and clinical use. Progress in brain research, vol 54. Elsevier, Amsterdam, pp 3-15

16 Caspers H, Speckmann EJ, Bingmann D, Lehmenkühler A (1986) Wirkung von CO_2 auf das Membranpotential einzelner Neurone. In: Grote J, Thews G (eds) Aktuelle Probleme der Atmungs- und Kreislaufregulation. Funktionsanalyse biologischer Systeme 15. Steiner, Stuttgart, pp 185-195

17 Caspers H, Speckmann EJ, Lehmenkühler A (1987) DC potentials of the cerebral cortex – sei-
    zure activity and changes in gas pressure. Rev Physiol Biochem Pharmacol 106: 127–178
18 Kersting U, Lehmenkühler A (1986) CO_2-induced decrease of potassium permeability across
    the blood-brain barrier. Neurosci Lett 28: S488
19 Kienecker EW, Knoche H, Bingmann D (1978) Functional properties of regenerating sinus
    nerve fibres in the rabbit. Neuroscience 3: 977–988
20 Lehmenkühler A (1979) Interrelationships between DC potentials, potassium activity, pO_2 and
    pCO_2 in the cerebral cortex of the rat. In: Speckmann EJ, Caspers H (eds) Origin of cerebral
    field potentials. Thieme, Stuttgart, pp 49–59
21 Lehmenkühler A, Bingmann D, Lange-Asschenfeldt H, Berges D (1978) Oxygen pressure and
    ictal activity in the cerebral cortex of artificially ventilated rats during exposure to oxygen high
    pressure. In: Silver IA, Erecinska M, Bicher HI (eds) Oxygen transport to tissue III. Plenum,
    New York, pp 679–685
22 Lehmenkühler A, Zidek W, Staschen M, Caspers H (1981) Cortical pH and pCa in relation to
    DC potential shifts during spreading depression and asphyxiation. In: Sykova E, Hnik P, Vyk-
    licky L (eds) Ion-selective microelectrodes and their use in excitable tissues. Plenum, New
    York, pp 225–229
23 Lehmenkühler A, Caspers H, Kersting U (1985) Relations between DC-potentials, extracellu-
    lar ion activities and extracellular volume fraction in the cerebral cortex with changes in pCO_2.
    In: Kessler M, Harrison DK, Höper J (eds) Ion measurements in physiology and medicine.
    Springer, Berlin Heidelberg New York, pp 199–205
24 Lehmenkühler A, Caspers H, Speckmann EJ, Bingmann D, Lipinski HG, Kersting U (1988)
    Neurons, glia and ions in hypoxia, hypercapnia and acidosis. In: Somjen GG (ed) Hypoxia
    and stroke. Plenum, New York
25 Lipinski HG, Bingmann D (1986) PO_2 dependent distribution of potassium in hippocampal
    slices of the guinea pig. Brain Res 380: 267–275
26 Lübbers DW (1968) The oxygen pressure field of the brain and its significance for the normal
    and critical oxygen supply of the brain. In: Lübbers DW et al. (eds) Oxygen transport in blood
    and tissue. Thieme, Stuttgart, pp 124–139   ·
27 Lux HD, Müller TH (1987) Calzium-abhängige Schrittmacherprozesse an der neuronalen
    Membran mit dem Zeitbedarf paroxysmaler Vorgänge. In: Speckmann EJ (ed) Epilepsie 86.
    Einhorn, Reinbek, pp 16–26
28 Speckmann EJ, Caspers H (1967) Les modifications du potentiel continu cortical pendant l'ar-
    rêt respiratoire. Rev Neurol (Paris) 117 (1): 5–19
29 Speckmann EJ, Caspers H (1969) Verschiebungen des corticalen Bestandpotentials bei Verän-
    derungen der Ventilationsgröße. Pflugers Arch 310: 235–250
30 Speckmann EJ, Caspers H (1974) The effect of O_2- and CO_2-tensions in the nervous tissue on
    neuronal activity and DC potentials. In: Remond A (ed) Handbook of electroencephalogra-
    phy and clinical neurophysiology, vol 2/C. Elsevier, Amsterdam, pp 71–89
31 Speckmann EJ, Caspers H (1975) Responses of spinal and cortical neurons to changes of pO_2
    and pCO_2 in blood and tissue. In: Purves MJ (ed) The peripheral arterial chemoreceptors.
    Cambridge University Press, London, pp 163–172
32 Speckmann EJ, Caspers H, Bingmann D (1973) Actions of hypoxia and hypercapnia on single
    mammalian neuron. In: Bicher H, Bruley DF (eds) Oxygen transport to tissue, vol 1. Plenum,
    New York, pp 245–250 (Advances in experimental medicine and biology, vol 37 A)
33 Zidek W, Lehmenkühler A, Caspers H, Lange-Asschenfeldt H (1978) Macromolecular buffer-
    ing reverses the CO_2 effect on the membrane potential in snail neurons. Pflugers Arch 377:
    R43

Possible Meaning of Different Ionic Changes in the Carotid Body During Hypoxia

M. A. Delpiano

Max-Planck-Institut für Systemphysiologie, Rheinlanddamm 201, 4600 Dortmund 1, FRG

Introduction

It is well known that the carotid body, a peripheral chemoreceptor located near the carotid sinus, is able to detect gas changes (PO_2, PCO_2/pH) in arterial blood composition and transduce them into nerve signals to regulate circulation and ventilation [19, 24].

Although it is not entirely understood how the essential mechanism for chemoreception functions, a lot of evidence hints at the idea that chemoreception in the carotid body comprises a sequential interaction of several events initiated by changes of the environmental PO_2, which induces glomus (type-I) cells to secrete transmitter(s) and depolarize nerve endings [16, 18]. According to this idea, chemoreception corresponds to a metabolically dependent stimulus-secretory response, which is closely related to the presence of extracellular Ca^{2+} and substrate, and where these chromaffin-like (type-I) cells storing and secreting transmitters may represent the primary receptor site.

The superfused carotid body in vitro is a good model to elucidate the intimate stimulus-secretory transduction mechanism dealing with the O_2 sensing function itself, since it is not influenced by either efferent nervous input nor microregulatory changes of local flow.

The aim of this article is to attempt an integrative interpretation of the different findings obtained from in vitro studies of the carotid body related to extracellular changes of the Ca^{2+} and K^+ activities as well as changes of the extracellular pH during hypoxia and to contribute a tentative model towards a better understanding of O_2 chemoreception.

Methods

Experiments were performed with carotid bodies (c.b.) excised from anesthetized cats and superfused in vitro in a leucite chamber (A), as illustrated in Fig. 1. The organs were superfused with a modified Locke's solution consisting of (mM): NaCl 128, KCl 5.6, $CaCl_2$ 2.1, D-glucose 5.5, $NaHCO_3$ 12, HEPES 7.0, adjusted with sucrose to 300 mosmol. The superfusion medium was equilibrated with dif-

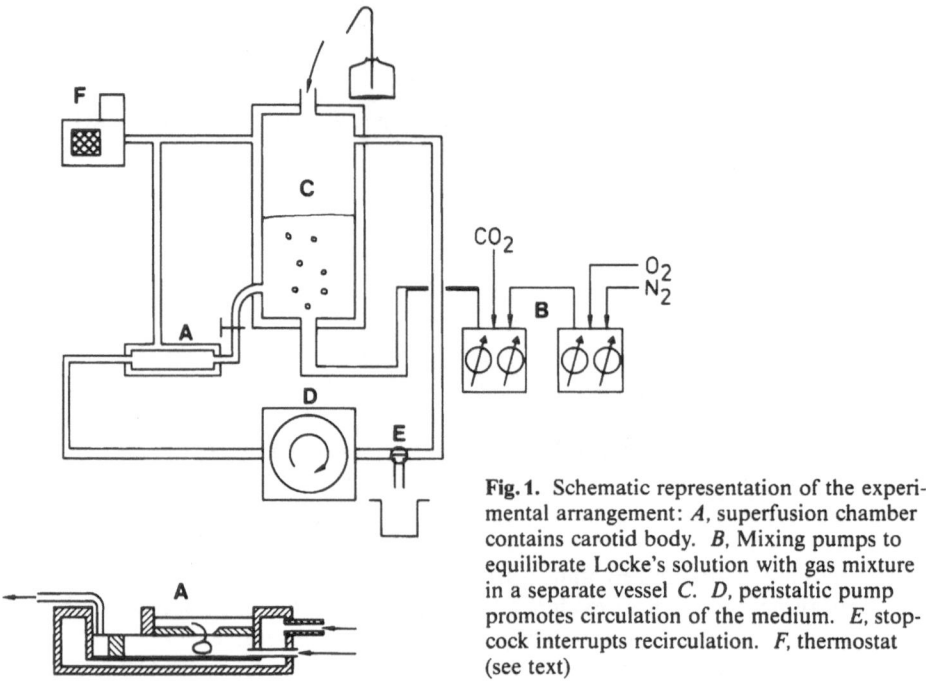

Fig. 1. Schematic representation of the experimental arrangement: *A*, superfusion chamber contains carotid body. *B*, Mixing pumps to equilibrate Locke's solution with gas mixture in a separate vessel *C*. *D*, peristaltic pump promotes circulation of the medium. *E*, stopcock interrupts recirculation. *F*, thermostat (see text)

ferent gas mixtures with the aid of gas mixing pumps (B) in separate vessels (C) connected to the superfusion chamber (A) by glass tubing. A peristaltic pump (D) drained the superfusate from the chamber (A) and returned it back to the reservoir (C). In some experiments, when drugs were used, the stopcock (E) was open to avoid recirculation. The whole system was thermostabilized with the aid of a thermostat (F). Temperature ($35°C$), pH (7.42 ± 0.02), and flow (3.6 ml/min) were maintained constant during the experiments.

The PO_2 in the superfusate (PmO_2) of about 188 ± 3 torr was referred to as normoxic and was continuously monitored with a PO_2 catheter electrode [27]. Hypoxia was created by decreasing the environmental PO_2 to about 25 ± 3 torr for 5 min, keeping the PcO_2 constant at about 25 ± 2 torr. The pH in the medium (pH_m) and the temperature were also controlled with pH electrodes (Ingold) and thermocouples (Sika), respectively.

The chemoreceptor discharge (nerv. resp.) of the cut end of the sinus nerve lifted into paraffin oil was recorded with two platinum electrodes, as reported elsewhere [8].

The extracellular Ca^{2+} and K^+ activities (aCa_e^{2+}, aK_e^+) and the extracellular pH (pH_e) were respectively measured with triple-barrelled ion-selective electrodes [15] and double-barrelled pH glass electrodes [11]. The electrodes were inserted into the c.b. tissue for about 400 μm by means of a step-motor driven manipulator (Nano-Stepper, WPF Instruments).

Results and Discussion

When the c.b. was stimulated by decreasing the environmental PO_2 from 188 to 25 torr, referred to as hypoxia, it could always be observed that together with the evoked chemoreceptor discharge other changes concerning the extracellular activity of Ca^{2+}, K^+, and pH also occurred.

Beginning with the analysis of the changes in aCa_e^{2+} and aK_e^+, it can be seen in Fig. 2a that lowering the PmO_2 under normal Ca^{2+} concentration in the medium (2.0 mM) always leads to a decrease in aCa_e^{2+} and to a biphasic increase in aK_e^+.

At the onset of hypoxia, the aCa_e^{2+} started immediately to decrease, almost at the same time the chemoreceptor discharge initiated its increase. The initial decrease in aCa_e^{2+} was followed by a small transient increase (Fig. 2a, arrow), which deflected to a clear decrease, when the PmO_2 reached its minimal value of about 25 torr (5 min).

Paralleling these aCa_e^{2+} changes, the aK_e^+ started to increase quickly (some seconds following the aCa_e^{2+} decrease) and reached its maximal peak of about 1.6 mM over the baseline (6.13 mM) 2 min before the PmO_2 had reached its lowest value. Afterwards, the maximal aK_e^+ increase, declined slowly at first and then more rapidly at the end of hypoxia, achieving an undershoot during the hypoxic recovering time (Fig. 2a). Both ionic changes, aCa_e^{2+} and aK_e^+, showed a delayed long-lasting return to initial values at the end of hypoxia.

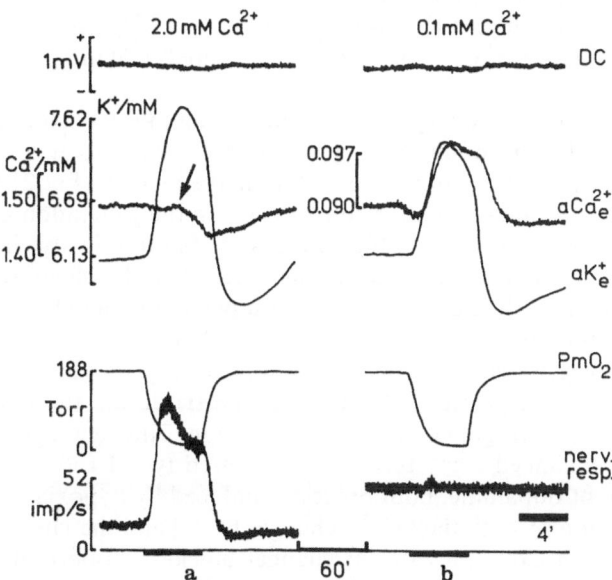

Fig. 2a, b. From *top* to *bottom:* bioelectrical potential (DC/mV), extracellular calcium activity (aCa_e^{2+}/mM), extracellular K^+ activity (aK_e^+/mM), oxygen partial pressure in the medium (PmO_2/Torr), and chemoreceptor discharge (nerv. resp./imp/s). **a** Changes in aCa_e^{2+}, aK_e^+, and nerve discharge during hypoxia at normal Ca^{2+} concentration in the medium (2.0 mM). **b** Reversal in aCa_e^{2+} and attenuation in aK_e^+, together with absence of chemoreceptor discharge to hypoxia, at low "critical" Ca^{2+} concentration (0.1 mM). The time, 4 min

By reducing the calcium concentration in the superfusate to a value of 0.1 mM, another pattern in the ionic response could be observed, as illustrated in Fig. 2b. When low calcium saline replaced normal saline, the extracellular aCa_e^{2+} deflected downward in a long-lasting form (not shown) to reach 12 min later a value of about 0.090 mM (Fig. 2b). Concomitant with this ionic change, the spontaneous nerve discharge increased, probably due to the destabilizing effect of low Ca^{2+} on the nerve membrane permeability [17]. In Fig. 2b, for convenience, the lowered baseline of aCa_e^{2+} (0.090 mM) was inserted at the same calibration level as the scale for normal calcium. Therefore, it is important to be careful in considering both different scales (logarithmic scale), which represent different decades of the calcium activity. Lowering PmO_2 at this low calcium concentration (0.1 mM), induced an initial aCa_e^{2+} decrease followed, however, by a clear and distinct increase. This increase in aCa_e^{2+} amounted to about 0.006 mM and returned to baseline without an undershoot when PmO_2 was restored (Fig. 2b). In other words, a decrease in the Ca^{2+} concentration of the superfusion medium always leads to a reversal of the aCa_e^{2+} response observed under normal conditions, probably because this procedure impairs the Ca^{2+} influx into the cells [23]. This increase in aCa_e^{2+} corresponded only to 20% of the aCa_e^{2+} changes (measured under normal calcium concentration, as indicated by the calibration scale in Fig. 2). The biphasic aK_e^+ changes induced by hypoxia were strongly attenuated at this low saline calcium, especially in their increase rather than in their undershoot (Fig. 2b). Attenuation of the aK_e^+ increase can be achieved by $CoCl_2$ (not shown), which also affects the aCa_e^{2+} decrease and chemoreceptor discharge [13]. It is suggested that the aK_e^+ increase represented in part K^+ efflux produced by Ca^{2+} influx and may signal, therefore, the existence of Ca^{2+}-activated K^+ channels [29] in glomus cells.

Concerning the chemosensory nerve response, it is interesting to establish that under these low "critical" calcium conditions hypoxia was not able to induce chemoreceptor discharge in spite of an enhanced resting nerve activity (Fig. 2b). These findings agree with the essential coupling role of Ca^{2+} in the generation of chemoreception reported by other authors [4, 16]. In the same figure it is also interesting to note that the ionic changes persisted in this case, when the chemoreceptor discharge was absent. Since the aCa_e^{2+} and aK_e^+ also persisted after chronic denervation of the carotid body [9] or nerve poisoning with tetrodotoxin (TTX) [10, 13], it suggested that they represent to a large extent ionic changes involved in the membrane permeability of the glomus cells (type-I, type-II) during the hormone release, and are not merely originated by nerve excitation. This idea agrees with findings reporting an enhanced Ca^{2+} turnover in cultured type-I cells, and the Ca^{2+}-dependent release of dopamine, both processes induced by hypoxia [16, 32]. Furthermore, the inhibition of both the aCa_e^{2+} changes [9, 13] and the chemoreceptor discharge (Fig. 3) with $CoCl_2$, a calcium channel blocker, supports this idea. Thus, looking at the nervous response evoked by hypoxia at different Co^{2+} and Ca^{2+} concentrations in Fig. 3, it can be seen that for a given Co^{2+} concentration (3 mM), the lower the Ca^{2+} concentration in the medium the stronger the Co^{2+} inhibition and vice versa. Although this is also true for aCa_e^{2+} changes, the chemoreceptor discharge is more sensitive to the effect of Ca^{2+} channel blockers [9, 13].

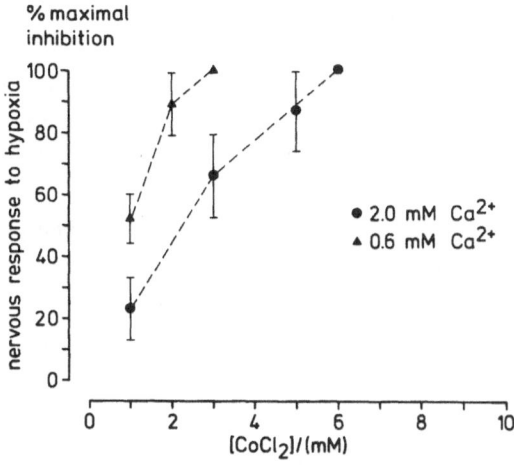

Fig. 3. Inhibitory effect of different CoCl$_2$ concentrations, in *abscissa*, on the chemosensory response to hypoxia, in *ordinate*, at two different Ca^{2+} concentrations in the medium.

These aCa$_e^{2+}$ changes may represent, to a great extent, influx and efflux of free Ca^{2+} occurring at glomus cells during hypoxia. Thus, under normal saline Ca^{2+} concentration the Ca^{2+} influx, i.e., decrease in aCa$_e^{2+}$, predominates over the efflux and masks it (Fig. 2a, arrows). However, when the saline Ca^{2+} concentration is lowered, the Ca^{2+} influx is more greatly reduced than is the Ca^{2+} efflux by this procedure [23], and only the small increase of aCa$_e^{2+}$ can be observed (Fig. 2b).

The aK$_e^+$ changes during hypoxia thus represent a composite response, as will be discussed below.

There is a link between the disturbance of the metabolic state of glomus cells and hormone release responsible for the chemoreceptor discharge, there is a fairly wide consensus that mitochondrial oxidative metabolism is directly involved in the initial step of the chemotransduction [3, 21]. However, so far, the cause of this energy control is not well defined and apparently chemoreception cannot be explained solely as a depression of oxidative metabolism [2]. Hence, the participation of the glycolytic pathway seems to be also involved [3, 11, 26, 37].

In the cat c.b. under normal glucose concentration (5.5 mM), hypoxia induced an acid shift in pH$_e$ together with the expected chemoreceptor discharge (Fig. 4). This tissue acidification initiated when the PmO$_2$ was still high and permitted good tissue oxygenation [8]. This particularity of the cat c.b. and the fact that tissue acidification could be prevented by zero glucose or inhibition of glycolysis leads to the assumption that the pH$_e$ decrease is associated with intracellular acidification by activation of glycolytic lactic acid production [11]. Moreover, as illustrated in Fig. 4, the magnitude of the pH$_e$ decrease is directly related to the stimulus intensity. For instance, a decrease in PmO^2 from 189 to 25 torr (corresponding to a strong hypoxia), led to a pH$_e$ change of about 0.06 units (Fig. 4, first hypoxia), while a decrease in PmO$_2$ to only 150 torr (corresponding to a mild hypoxia), produced a pH$_e$ change of only about 0.02 units (Fig. 4, second hypoxia). On the other hand, increasing PmO$_2$ (hyperoxia) induced a slight increase in pH$_e$ and inhibited in each case the chemoreceptor discharge. A similar alkaline shift has been observed in the cat c.b. during hypoxia without glucose [11]. The origin of this alka-

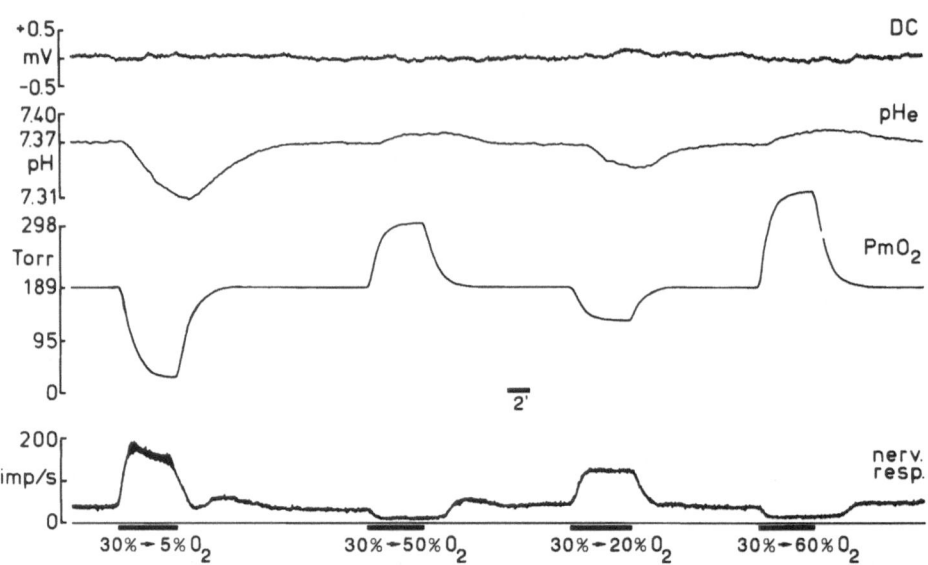

Fig. 4. Chemoreceptor response and changes in pH_e during hypoxia and hyperoxia under normal glucose (5.5 mM) of the cat c.b. in vitro. From *top* to *bottom:* bioelectrical potential (DC/mM), extracellular pH (pH_e/pH units), oxygen partial pressure in the medium (PmO_2/torr), and nerve discharge during hypoxia and hyperoxia (nerv. resp./imp/s)

line shift may be related to some metabolic production of anions, possibly HCO_3^-, HPO_4^-, or also Cl^-/HCO_3^- cotransport, to compensate an acid shift and establish electroneutrality [25].

These considerations suggest that pH_e decrease produced by hypoxia may have arisen from H^+ transport out of the cell in order to maintain the intracellular acid-base balance disturbed by the glycolytic lactic acid production [11]. Since this acid shift of pH_e can be prevented by Na^+-free saline or ouabain during hypoxia but not during anoxia, it is suggested that two mechanisms contribute to H^+ ion extrusion in the cat c.b.; the former produced by the Na^+/H^+ exchanger and the latter by a passive lactic acid diffusion [7]. The question arose as to how the glycolytic pathway is linked to the genesis of the chemoreceptor discharge. The fact that during hypoxia the pH_e started to decrease at a time when the chemoreceptor discharge had just reached its maximal increase (Fig. 4) provided clues suggesting more a regulatory role of glycolysis on chemotransduction rather than a determining one. This assumption is supported by the restoring effect of pyruvate upon the diminished chemosensory response under zero glucose or 2-deoxyglucose [11, 12]. As shown in Fig. 5, zero glucose produced attenuation of both the acid pH_e shift and chemoreceptor response to hypoxia in a time-dependent mode. Thus, 100 min after superfusion without glucose (■), the ΔpH_e was strongly reduced showing a slight alkaline shift at the end of hypoxia (Fig. 5 a). The chemoreceptor discharge was reduced by 80% after this time (Fig. 5 b). However, addition of equimolar pyruvate concentration (◆) 200 min after superfusion without glucose restored almost completely the chemoreceptor discharge, but did not restore the inhibited pH_e de-

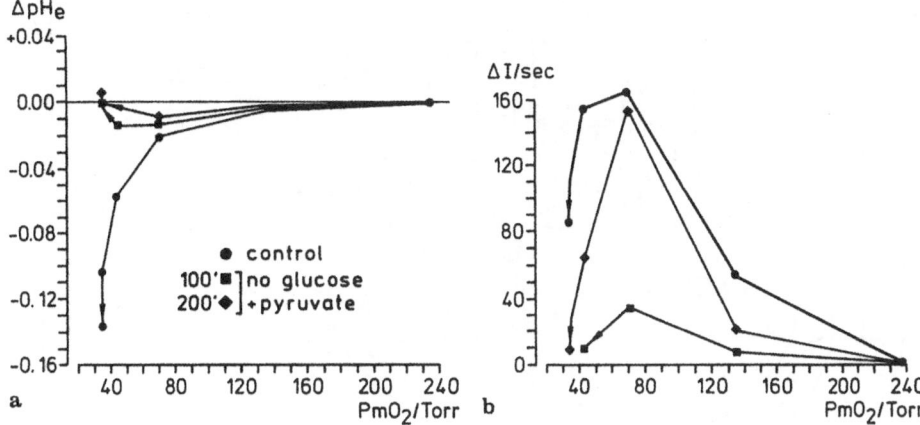

Fig. 5a, b. Changes in extracellular pH and nerve discharge under different substrate conditions. **a** pH_e during hypoxia with normal glucose ($\bullet$), after 100' superfusion with zero glucose ($\blacksquare$), and addition of pyruvate after 200' with zero glucose ($\blacklozenge$). **b** The same substrate conditions are shown for the chemosensory response

crease (Fig. 5a, b). Taking into account that the pyruvate effect could be exerted, even if the glycolysis was inhibited [12], and that malonate, an inhibitor of the citric acid cycle [35], competitively inhibited the pyruvate effect [7], it can be presumed that this mechanism is closely associated to substrate activations of the oxidative metabolism by means of pyruvate oxidation in the citric acid cycle, providing increased flow of reducing equivalents for the mitochondrial respiratory chain. This assumption is supported by fluorometric studies of the NADH fluorescence during hypoxia [30]. Hypoxia caused biphasic NADH fluorescence changes consisting in an initial decrease followed by an increase which indicated oxidation and reduction of NADH and NAD^+, respectively. Pyruvate, in the absence of exogenous glucose during hypoxia, enhanced the area enclosed in the initial downward deflection (oxidation of NADH) signalizing activation of oxidative metabolism [20], but did not restore the increase in NADH fluorescence (reduction of NAD^+) prevented under zero glucose and, therefore, attributed to the activation of glycolysis [30]. To conciliate the simultaneous reduction of cytochrome a_3 occurring during hypoxia, that signals depression of the oxidative metabolism, with the oxidation of NADH, that signals activation of the oxidative metabolism, Mills and Jöbsis [20] assumed the existence of two different respiratory chains: one with an unusually low oxygen affinity cytochrome a_3 undergoing reduction at high PO_2 values and thereby intimately linked to a generation of the chemoreceptor discharge, while another, probably located in a different cell, possesses an unusually high oxygen affinity cytochrome a_3, remains in the oxidized state at very low PO_2 values (7–9 torr), and mediates oxidation of NADH to maintain energy requirements. Although this hypothesis has been questioned by other authors [1] and the existence of two different respiratory chains could be unnecessary in as much as common mitochondrial respiratory chains can also be reduced by decreasing PO_2 within physiological range [36], these metabolic studies [30] for the first time, pro-

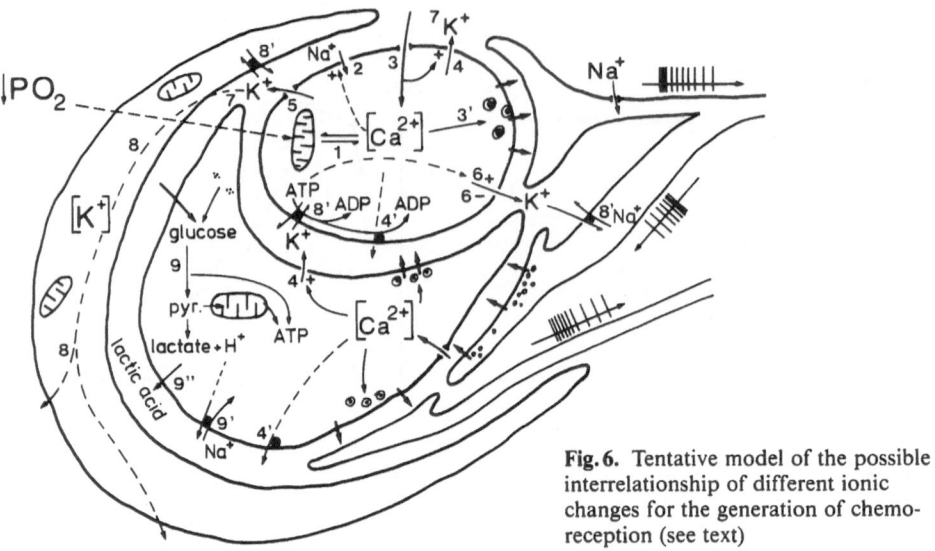

Fig. 6. Tentative model of the possible interrelationship of different ionic changes for the generation of chemoreception (see text)

vided clues for a more complex metabolic interaction for inducing chemosensory response.

In conclusion, these changes in aCa_e^{2+}, aK_e^+, and pH measured during hypoxia might be involved in some manner in the generation and maintenance of the chemoreceptor discharge. Hence, in seeking a tentative model that integrates these different mechanisms for sensing PO_2 in the c.b., it should be noted, as schematically illustrated with different steps *(1–9″)* in Fig. 6, that probably the first event arising with the onset of hypoxia is *(1)* an intracellular liberation of Ca^{2+} from mitochondria due to the decreased PO_2-dependent oxidative metabolism [30]. In this respect, it is reported that deterioration of mitochondrial Ca^{2+} uptake in the c.b. by La^{3+} or ruthenium red increases chemoreceptor discharge [33]. This initial Ca^{2+} increase, however, should not be sufficient for induced nerve excitation, since low "critical" Ca^{2+} in the medium prevents the chemoreceptor discharge (Fig. 2b), but does not prevent cAMP increase, which is dependent of cytosolic Ca^{2+} [14]. Therefore, the cytosolic Ca^{2+} increase could depolarize the cell by gating *(2)* "nonselective cation channels" [28], and *(3)* lead to a Ca^{2+} influx, *(3′)* essential for the transmitter release and the nerve excitation. Later on, when cytosolic Ca^{2+} reaches threshold, the K^+ permeability increases due to an activation of *(4)* 'Ca^{2+}-activated K^+ channels' [29], as assumed from the aK_e^+ increase during hypoxia since low saline Ca^{2+} (Fig. 2b) or Ca^{2+} channel blockers [10] attenuate it. Moreover, threshold Ca^{2+} may activate *(4′)* a Ca^{2+}-ATPase, since a Ca^{2+}-ATPase has been suggested in glomus cells [34] and because the unmasked increase in aCa_e^{2+} during low Ca^{2+} concentration (Fig. 2b) can be blocked by inhibitors of the Ca^{2+}-ATPase [13]. Voltage-dependent K^+ channels gated during the depolarization *(5)* may also contribute to the increase in aK_e^+, since TEA inhibits this increase [9]. Here it should be noted that *(6)* ATP probably also regulates K^+ permeability, because metabolism of the c.b. [30] as well as the chemoreceptor discharge

(Fig. 5) are strongly dependent on supply of exogenous glucose, and ATP-sensitive K^+ channels have been demonstrated in cells whose electrical activity is influenced by glucose and metabolic poisons [5, 31].

So far, nothing is known about intracellular ATP interactions with K^+ permeability in the c.b., merely that exogenous ATP increases the chemoreceptor discharge [22]. However, keeping in mind the dual behavior of the oxidative metabolism during hypoxia [30], it is not unreasonable to expect that such glomus (type-I) cells, which initially depolarize due to a depression of the oxidative metabolism, become hyperpolarized afterwards, when the ATP decrease *(6+)* gates K^+ channels increasing K^+ permeability, while other glomus (type-I) cells, undergoing activation of metabolism with increased ATP production become depolarized when *(6−)* ATP blocks the K^+ channel decreasing K^+ permeability.

These increases in K^+ permeability produced by these different mechanisms (Ca^{2+}-, voltage-, and ATP-sensitive K^+ channels) that follow the initial depolarization may induce secondary processes: (a) limitation of neurosecretion due to cell hyperpolarization, (b) modulation of nerve excitability, and probably *(7)* secondary depolarization due to extracellular K^+ accumulation, and from this (c) activated K^+ clearance by means of *(8)* passive glia (type-II) cell K^+ re-uptake and *(8')* active K^+ transport in different cells, as assumed from the ouabain-sensitive aK_e^+ undershoot [10].

At least, during hypoxia, *(9)* glycolysis is activated, probably contributing as an additional energy source in those cells possessing a PO_2-dependent O_2 consumption, whereas intracellular acidification can be regulated *(9')* by the Na^+/H^+ exchanger and *(9'')* lactic acid diffusion. However, it cannot be ruled out that glycolysis may be also involved in regulating the ionic permeability, since intracellular acidification inhibits "Ca^{2+}-activated K^+ channels" [6] and so counteracts hyperpolarization. This complex alternation of cell excitability with depolarization and hyperpolarization develops in an oscillatory form until PO_2 deficiency is normalized – and may represent the link between metabolism and cell excitability for chemoreception.

References

1 Acker H, Eyzaguirre C, Goldman WF (1985) Redox changes in the mouse carotid body during hypoxia. Brain Res 330: 158–163
2 Acker H, Eyzaguirre C (1987) Light absorbance changes in the mouse carotid body during hypoxia and cyanide poisoning. Brain Res 409: 380–385
3 Anichkov SV, Belenkii ML (1963) Pharmacology of the carotid body chemoreceptors. Pergamon, Oxford, p 164
4 Bernon R, Leitner LM, Roumy M, Verna A (1983) Effects of ion containing liposomes upon the chemoafferent activity of the rabbit carotid body superfused in vitro. Neursci Lett 35: 289–295
5 Cook DL, Hales CN (1984) Intracellular ATP directly blocks K^+ channels in pancreatic B-cells. Nature 311: 271–273
6 Cook DL, Masatoshi I, Fujimoto WY (1984) Lowering of pHi inhibits Ca^{2+}-activated K^+ channels in pancreatic B-cells. Nature 311: 269–271
7 Delpiano MA (1987) Glycolysis as a link for chemoreception? In: Ribeiro JA, Pallot DJ (eds) Chemoreceptors in respiratory control. Croom Helm, London, p 59

8 Delpiano MA, Acker H (1980) Relationship between tissue PO_2 and chemoreceptor activity of the cat carotid body in vitro. Brain Res 195: 85-93

9 Delpiano MA, Acker H (1984) The extracellular Ca^{++} and K^+ activities in the cat carotid body in vitro and their relationship to chemoreception. In: Pallot DJ (ed) The peripheral arterial chemoreceptors. Crooms Helm, London, p 101

10 Delpiano MA, Acker H (1984) Simultaneous response of the extracellular Ca^{++} and K^+ activity during hypoxia and hypercapnia and their possible interdependence in the superfused cat carotid body. Pflugers Arch [Suppl] 402: R35

11 Delpiano MA, Acker H (1985) Extracellular pH changes in the superfused cat carotid body during hypoxia and hypercapnia. Brain Res 342: 273-280

12 Delpiano MA, Acker H (1985) Extracellular pH responses to different stimuli in the superfused cat carotid body. Adv Exp Med Biol 191: 709-717

13 Delpiano MA, Acker H (1988) The extracellular response of the aCa_e^{2+} and aK_e^+ during hypoxia and hypercapnia of the cat carotid body in vitro. Brain Res (to be published)

14 Delpiano MA, Starlinger H, Acker H (1985) Changes in the cAMP content of the superfused cat carotid body produced by initial PO_2 decrease in the medium. Pflugers Arch [Suppl] 405: R37

15 Dufau E, Acker H, Sylvester D (1982) Triple-barrelled ion-sensitive microelectrodes for simultaneous measurements of two extracellular ion activities. Med Prog Technol 9: 33-38

16 Fidone S, Gonzalez C, Yoshizaki K (1982) Effects of low oxygen on the release of dopamine from the rabbit carotid body in vitro. J Physiol (Lond) 333: 93-110

17 Frankenhäuser B, Hodgkin AL (1957) The action of calcium on the electrical properties of squid axons. J Physiol (Lond) 137: 218-244

18 Hayashida Y, Koyano H, Eyzaguirre (1980) An intracellular study of chemosensory fibers and endings. J Neurophysiol 44: 1077-1088

19 Heymans C, Neil E (1958) Reflexogenic areas of the cardiovascular system. Churchill, London, p 153-184

20 Jöbsis FF, Duffield JC (1967) Oxidative and glycolytic recovery metabolism in muscle. J Gen Physiol 50: 1009-1047

21 Joels N, Neil E (1963) The excitation mechanism of the carotid body. Br Med Bull 19: 21-24

22 Joels N, Neil E (1968) The idea of a sensory transmitter. In: Torrance RW (ed) Arterial chemoreceptors. Blackwell, Oxford, p 153

23 Kondo S, Schulz I (1976) Ca^{2+} fluxes in isolated cells of rat pancreas. Effect of secretagogues at different Ca^{2+} concentrations. J Membr Biol 29: 185-203

24 Korner PI (1965) The role of arterial chemoreceptors and baroreceptors in the circulatory response to hypoxia of the rabbit. J Physiol (Lond) 180: 279-303

25 Kraig RP, Ferreira-Filho CR, Nicholson C (1983) Alkaline and acid transients in cerebellar microenvironment. J Neurophysiol 49: 831-850

26 Landgren S, Liljestrand G, Zotterman Y (1954) Impulse activity in the carotid sinus nerve following intracarotid injections of sodium-iodo-acetate, histamine hydrochloride, lergitin, and some purine and barbituric acid derivates. Acta Physiol Scand 30: 149-160

27 Lübbers DW, Baumgärtl H, Fabel H, Huch A, Kessler M, Kunze K, Riemann H, Seiler D, Schuchhardt S (1969) Principles of construction and application of various platinum electrodes. Prog Respir Res 3: 136-146

28 Maruyama Y, Petersen OH (1984) Single calcium-dependent cation channels in mouse pancreatic acinar cells. J Membr Biol 81: 83-87

29 Meech RW, Standen NB (1975) Potassium activation in Helix Aspersa neurons under voltage clamp: a component mediated by calcium influx. J Physiol (Lond) 249: 211-239

30 Mills E, Jöbsis FF (1972) Mitochondrial respiration chain of carotid body and chemoreceptor response to changes in oxygen tension. J Neurophysiol 35: 405-428

31 Noma A (1983) ATP-regulated K^+ channels in cardiac muscle. Nature 305: 147-148

32 Pietruschka F (1985) Calcium influx in cultured carotid body cells is stimulated by acetylcholine and hypoxia. Brain Res 347: 140-143

33 Roumy M, Leitner LM (1977) Role of calcium ions in the mechanism of arterial chemoreceptor excitations. In: Acker H, Fidone S, Pallot DJ, Eyzaguirre C, Lübbers DW, Torrance RW (eds) Chemoreception in the carotid body. Springer, Berlin Heidelberg New York, p 257

34 Starlinger H (1982) ATPases of the cat carotid body and of the neighbouring ganglia. Z Naturforsch [c] 37: 532–539
35 Webb JL (1966) Enzyme and metabolic inhibitors. Academic, New York, p 1
36 Wilson DF, Erecinśka M, Drown C, Silver IA (1979) The oxygen dependence of cellular energy metabolism. Arch Biochem Biophys 195: 485–493
37 Winder CV (1937) On the mechanism of stimulation of carotid gland chemoreceptors. Am J
Physiol 118: 389–398

Oxygen and Glycolysis in the Retina of the Compound Eye of a Crab

H. Langer[1], M. Delpiano[2], U. Knollmann[1], and H. Acker[2]

[1] Institut für Tierphysiologie, Ruhr-Universität Bochum, 4630 Bochum, FRG
[2] Max-Planck-Institut für Systemphysiologie, Rheinlanddamm 201, 4600 Dortmund 1, FRG

Introduction

The retina is – like other parts of the nervous system – a tissue with a high demand for oxygen. There is a very good supply in vertebrate retinas (from both sides of the retina in mammals) and a tracheol for each ommatidium in insect eyes. Nevertheless, under normal physiological conditions in vivo, there is a production of lactate in the retina of mammals, as was shown for the first time by Warburg and coworkers in 1924 (Warburg et al. 1924; for review see Sickel 1972). This aerobic glycolysis may cause a higher partial pressure of free oxygen in the retina by liberating oxygen from the hemoglobin in dependence on CO_2 and pH (Bohr effect). On the other hand, the metabolism of the insect retina is purely aerobic (Autrum and Tscharntke 1962); Hamdorf and Kaschef 1964; Hamdorf and Langer 1966; Tsacopoulos et al. 1981), and production of lactate was found only under lack of oxygen, when the specific function (as measured by electrophysiological methods) is already abolished (Hoffmann 1960; Langer 1962).

The higher crustaceans (Malacostraca) are a group of arthropods, in which oxygen supply to the tissue is – in contrast to insects – made by the hemolymph containing hemocyanin as oxygen carrier. Most of them have well-developed compound eyes with structures nearly related to those of insects. Crustacean eyes are supplied by the hemolymph which is unidirectionally streaming through defined lacunas within the eye-stalk. They consist of several thousand units – ommatidia – with seven or eight visual cells and their own dioptric apparatuses, separated from each other by so-called pigment cells, which ensheath the receptors, and by large extracellular spaces (Figs. 1 a, 2). The retina tissue is separated from the hemolymph only by a basement lamina.

For the investigation of the metabolism in a crustacean retina, we used the eyes of a terrestrial shore-crab, *Ocypode ryderi,* because of its prominent eye and its habitat and behavior (Henning and Langer 1986): the eye is a nearly complete cylinder consisting of about 30000 ommatidia which are very thin, but about 0.5 mm long, providing a considerable amount of retina tissue (about 10 mg dry retina material per eye). This eye has to deal with an extremely wide range of light intensities under its normal biological conditions. The activity pattern of this crab is controlled by the tides, independent of the daily light and dark cycle. There are light intensities of > 150000 lux in bright daylight and < 0.01 lux on moonless nights. Therefore, its retina should be a very suitable object for a study of the energy-releasing metabolism under different physiological conditions.

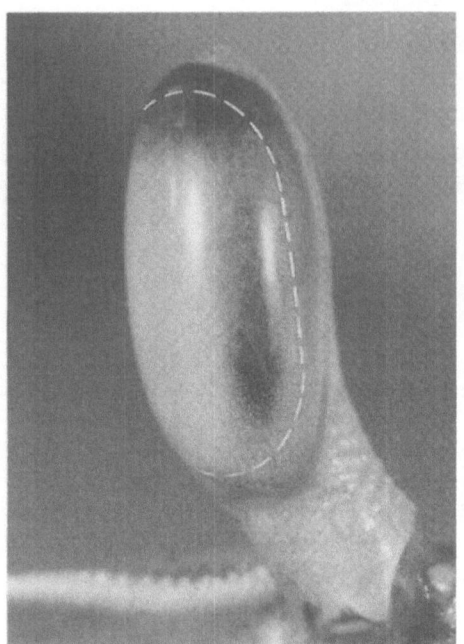

Fig.1. a Frontal view of a right eye-stalk of *Ocypode ryderi* in normal upright position; *broken white line* marks incision for preparation. **b** Eye preparation in superfusion vessel with two electrodes: PO_2/DC double-barreled microelectrode (right) in the tissue and PO_2 electrode (left) in the bath solution

Methods

From the compound eye an elliptical, bowl-shaped piece was cut out of the frontolateral area by means of a razor blade, the sectional plane was situated proximal to the basal lamina and perpendicular to the axes of the ommatidia. In Fig. 1 a, the dotted line indicates the edge of the intersection. The neuronal tissue proximal of the basal lamina was removed. This preparation was directly used for measurements of enzyme activities and – after shock freezing in liquid nitrogen and freeze drying – for analyses of substrate and metabolite concentrations. The same preparation was used for superfusion and incubation experiments in closed and open vessels, for determination of pH and oxygen contents in tissue, and thereafter for measurements of changes in metabolite concentrations.

Results

As an indicator for the capacity of metabolic pathways, the specific activities of several enzymes acting in energy-releasing metabolism were measured in the eyes of *Ocypode* and the retina of the crayfish *Astacus leptodactylus* and – for comparison – in the eyes of a fly, *Calliphora erythrocephala* (Rivera and Langer 1983 a, b), which is of the same morphological type (apposition eye) as that of *Ocypode*. The results (some of them given in Table 1) indicate that in *Ocypode* there is a very high capacity of the glycolytic pathway (LDH) and a relatively low capacity of

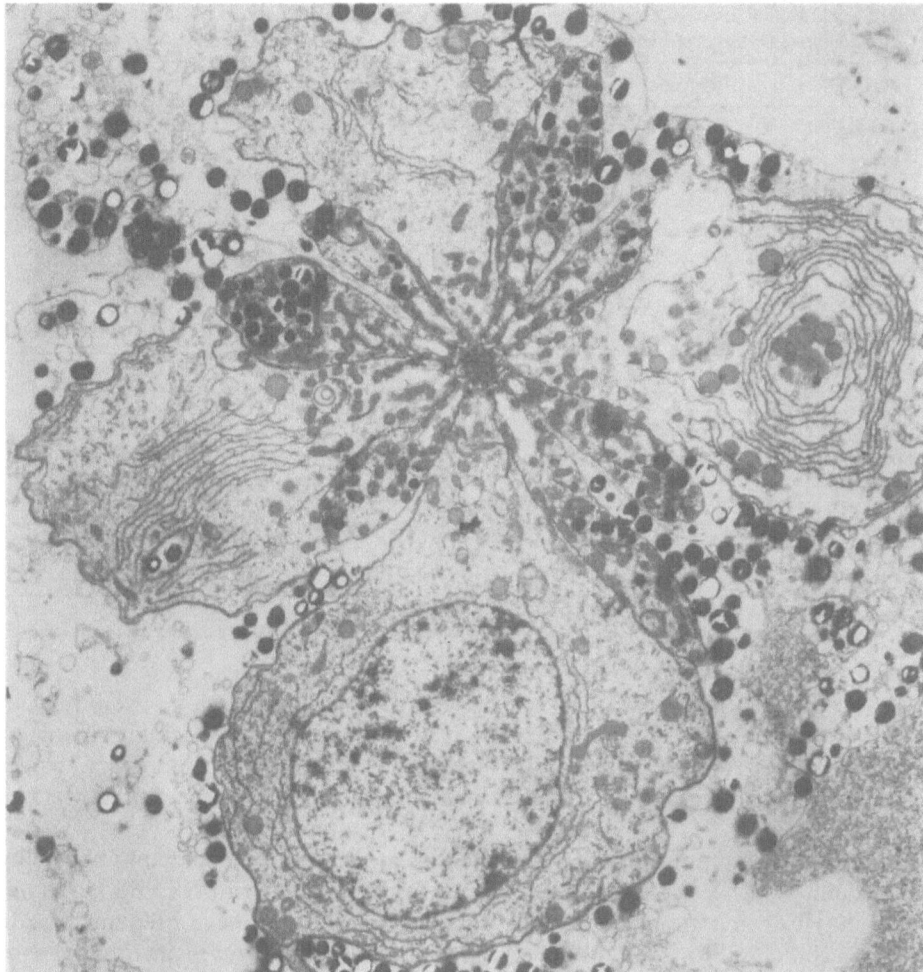

Fig. 2. Electron micrograph of a cross section through the most distal part of an ommatidium of *Ocypode ryderi* at the level of the nucleus of the short eighth receptor cell. This cell consists of four lobes (without pigment granules), one of them containing the nucleus; microvilli of all four lobes constitute the central rhabdom. Small distal parts of the seven long receptor cells (with dark pigment granules) are situated between the four lobes. The rhabdomeres of these cells (not seen in this section) build the rhabdom in most of its length. The receptor cells are surrounded by so-called pigment cells. The ommatidia are separated from each other by large extracellular spaces. Note numerous mitochondria surrounding the rhabdomere; they are concentrated in the most distal parts of all receptor cells. Magnification 5000×. Preparation and photography by Dr. U. Henning, Bochun

oxydative degradation (CS) as compared with the eye of *Calliphora*. The exclusively aerobic metabolism of the insect eye is correlated with a relatively low concentration of the phosphagen, arginine phosphate. The relation ArgP:ATP is about 1 in the eye of the fly, but about 4 in that of *Ocypode* (and also in the hermit crab *Eupagurus;* Langer et al. 1976), and still higher in the retina of the superposition eye of the crayfish *Astacus* (Lues 1978). In untreated eyes of *Ocypode*, the val-

Table 1. Maximal specific activities of enzymes in eyes of *Ocypode ryderi, Astacus leptodactylus,* and *Calliphora erythrocephala.* Data are given in mU × mg total protein^{-1} ± S. D.; $n = 8$

	Ocypode (eye)	*Astacus* (retina)	*Calliphora* (eye)
PH (total)[a]	31.1 ± 4.4	44.1 ± 8.2	12.7 ± 2.4
HK	4.7 ± 1.0	34.0 ± 6.9	16.8 ± 1.8
PFK	1.5 ± 0.1	4.3 ± 0.4	4.7 ± 1.4
GAPDH	169 ± 14	468 ± 41	1010 ± 94
PK	161 ± 19	639 ± 94	966 ± 97
G6PDH	24.0 ± 4.1	3.3 ± 0.8	6.5 ± 1.4
LDH	342 ± 50	382 ± 32	52.8 ± 7.6
GDH	11.6 ± 3.3	6.8 ± 1.8	61.3 ± 9.6
GPox	1.6 ± 0.5	0.9 ± 0.1	4.5 ± 0.4
CS	1.8 ± 0.6	12.7 ± 2.0	268 ± 41
IDH$_D$	0.3 ± 0.1	0.5 ± 0.3	22.9 ± 4.3
IDH$_P$	16.8 ± 2.2	13.2 ± 2.3	66.2 ± 9.0
MDH	881 ± 101	1926 ± 345	3150 ± 480
CCox	6.3 ± 1.7	8.2 ± 2.5	391 ± 50
Mg-ATPase	26.1 ± 3.2	122 ± 5	186 ± 24
NaK-ATPase	4.2 ± 1.4	10.9 ± 2.7	45.0 ± 6.7
ArgK	3470 ± 340	14080 ± 820	1300 ± 120

[a] Phosphorylase A represents about 40% of the total in each species

ue of the energy charge is high (0.85–0.90); unusually, the concentration of AMP is always higher than that of ADP (Table 2). During illumination – very high intensity during the day or lower intensity during the night – the arginine phosphate is remarkably reduced, while the ATP content is only slightly diminished. Nevertheless, the energy charge is reduced, mainly because of an increase in AMP concentration.

Oxygen consumption was investigated on preparations incubated in physiological saline (Cole 1941), modified by replacing boric acid with HEPES (10 mmol) and NaHCO$_3$ (6 mmol), pH 7.5. Under this condition, the eye preparation consumed about 2 µl O$_2$ × mg protein^{-1} × h^{-1} at 25° C. This figure is nearly the same as found earlier in the isolated retina from the superposition eye of the crayfish *Astacus* (Lues 1978).

During incubation, lactate is produced by the retina, while sufficient oxygen is present in the incubation medium (with a higher partial pressure than provided by the hemocyanin of the hemolymph in vivo). A comparable lactate production is well known from the mammalian retina (Warburg et al. 1924), while insect eyes do not produce any lactate as long as oxygen is available. Using incubation or superfusion methods for measuring oxygen consumption, there still remained the problem that the lactate production might result from insufficient supply of oxygen from the medium to the distal parts of the retina where most of the mitochondria of the visual cells are situated. Therefore, measurements of oxygen pressure were carried out with a PO$_2$ microelectrode in different depths of the retina. A pH/DC or PO$_2$/DC double-barrelled electrode (Acker et al. 1983) was introduced step by step into the retina by impaling the basal lamina, while avoiding to hit the projecting ends of the axons. The preparation (Fig. 1 b) was superfused at a flow rate of 4 ml × min^{-1} with the physiological saline (see above), supplemented by 5.5 m*M*

Table 2. Contents of adenine nucleotides and phosphagen and calculated energy charge in the retina of *Ocypode ryderi*. One eye frozen in liquid nitrogen immediately after removal from the animal and dissected after freeze-drying served as control. From the second eye of the same animal a preparation was made (see Methods) and incubated for 2 to 5 h in an open flow system (25° C, flow rate $1-4$ ml $\times$ h^{-1}). After this incubation the preparation was frozen and treated like the controls. Values are given in nmol $\times$ mg dry weight^{-1} $\pm$ S. D.; $n = 6$

	Control retina	Incubated retina
Arginine phosphate	24.1 ± 4.1	19.8 ± 4.8
ATP	6.1 ± 1.6	5.7 ± 1.4
ADP	0.4 ± 0.3	0.5 ± 0.3
AMP	1.3 ± 1.0	2.3 ± 1.7
Energy charge	0.85	0.71

Table 3. Contents of glycogen, glucose, lactate in and release of lactate from the retina of *Ocypode ryderi*. Treatment of controls and incubated preparations as described in Table 2. Values are given in nmol $\times$ dry weight^{-1} $\pm$ S. D.; $n = 6$

	Control retina	Incubated retina
Glycogen	659 ± 197	666 ± 242
Glucose	21.8 ± 6.8	30.7 ± 18.0
Lactate	2.4 ± 2.3	18.9 ± 6.5
Rate of lactate release (nmol $\times$ mg dry weight^{-1} $\times$ h^{-1})		6.6 ± 1.9

glucose. The pH of this medium was adjusted to 7.3 in an equilibration reservoir with CO_2 (approx. 3%) which was continuously bubbled through the solution in a gas mixture with O_2 (20% or 40%, respectively) and N_2 (for details, see the contribution by Delpiano in this volume, p. 193). The electrode was mainly situated between the ommatidial cells in the large extracellular space, which comprises most of the volume of the retina.

Moving the electrode stepwise through the superfusion medium to the basal lamina, a decrease in oxygen pressure was found already at a distance of 200 μm away from the tissue surface; after impaling the basal lamina, this decrease continued over the next 200 μm and then reached a constant level depending on the oxygen pressure of the superfusing solution (Fig. 3).

With respect to the lactate production, measurements of pH within the retina are of great interest (Fig. 4). At about 200 μm above the tissue surface, the pH of the incubation fluid (7.3) was still found. Nearer to the retina, the pH decreased down to about 100 μm within the tissue and remained constant at around 7.17. Within the tissue, this pH – lower as compared with the bath-fluid – remained constant, which points to a very efficient buffering within the tissue.

In further experiments, the oxygen partial pressure in the bath-fluid was reduced to about 150 torr (Fig. 3), which is nearer to the physiological conditions, but still more than in the hemolymph (the hemocyanin of *Ocypode* has a P_{50} value of about 12 torr at pH 7.8 and 30° C; Morris and Bridges 1985). The O_2 pressure

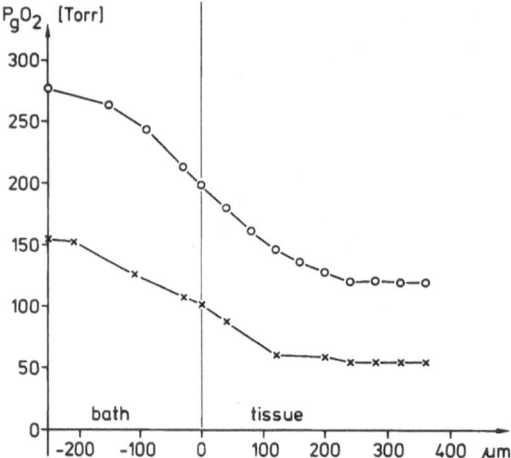

Fig. 3. PO₂ profile within the superfused retina of *Ocypode ryderi*. PO₂ profiles were measured by PO₂ microelectrodes under two different PO₂ situations in the superfusing medium (○, 280 torr; ×, 150 torr). The depth of puncture is given on the *abscissa*, the PO₂ on the *ordinate*. 0 μm represents the proximal surface of the basal lamina covering the retina; negative values represent the distance of the position of the electrode tip from this level in the bath, while positive values represent distance from this level in the tissue

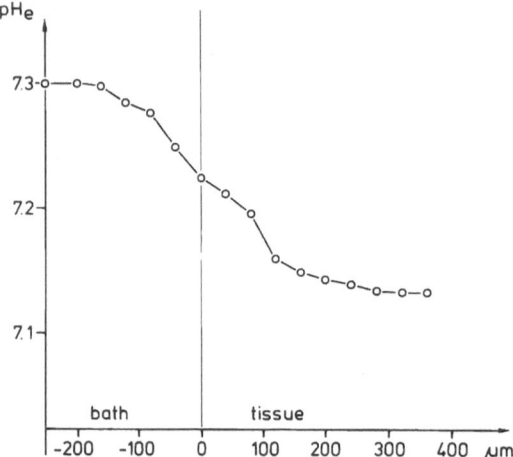

Fig. 4. pH profile within the superfused retina of *Ocypode ryderi* measured by a pH-sensitive microelectrode at 280 torr PO₂. Details of diagram as in Fig. 3

changes in nearly the same course in the range between 200 μm above and below the tissue surface; also in this case the O_2 pressure does not decrease to zero.

To investigate the effects of anoxia, the PO₂ was lowered in the superfusion medium to less than 20 torr within 3 min (Fig. 5). Thereby, the tissue PO₂ at about one-half of the length of the ommatidia (approx. 280 μm) reached zero level after about 6 min. The pH value did not change under these conditions at all. Later on, oxygen was supplied again and normal partial pressure was established in the tissue within less than 10 min.

The same preparation, mounted in an open-flow system, was used for investigating lactate production under high oxygen supply. The incubation was made at 25° C with a flow rate of 1 to 4 ml × h⁻¹ of air-saturated physiological saline (approx. 150 torr O_2). At up to 5 h of incubation, the tissue remained intact and the energy charge was only slightly reduced, mainly because of increase of AMP. Arginine phosphate was diminished by about 20% (Table 2); the contents of glycogen and glucose were not significantly changed. Starting after 45 min and during the entire rest of the incubation time, a continuous release of lactate at a constant

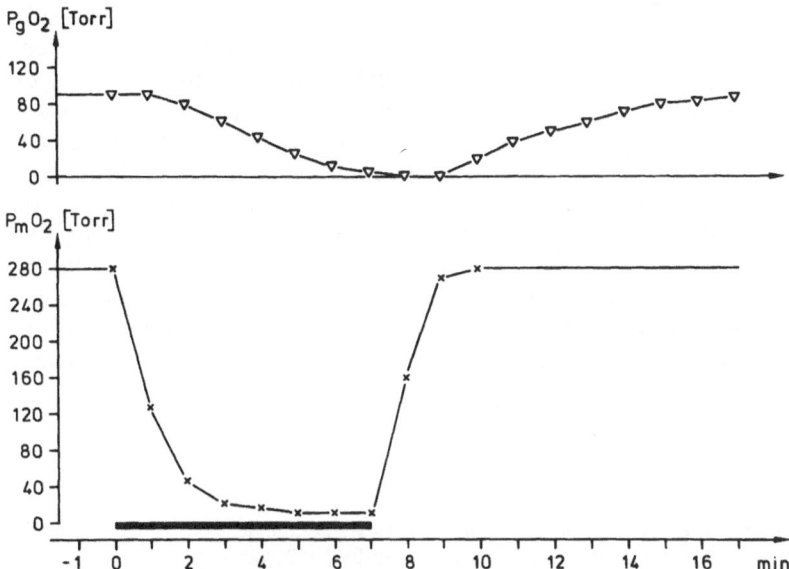

Fig. 5. Time course of tissue PO_2 in the retina of *Ocypode ryderi* during changes of the PO_2 in the superfusion medium. The *upper part* of the figure shows the PO_2 measured by a PO_2 electrode within the retina tissue in 280 μm depth, while the *lower part* gives the PO_2 of the medium (× = PmO_2) as measured using a catheter electrode. The anoxic exposure time is marked by the *horizontal bar*

rate occurred (Table 3); at the end of incubation the actual concentration of lactate was found increased by 800% within the tissue. In earlier experiments with incubations for 30 min in a closed system, the same rate of lactate production was found under aerobic conditions (no difference between media saturated with air or pure oxygen); anaerobic incubation resulted in a rate of lactate production about threefold higher. Evidently, aerobic incubation does not need the full capacity of lactate formation in the retina tissue.

Discussion

These experiments demonstrate the occurrence of aerobic glycolysis under physiological conditions within the retina tissue of a malacostracic crustacean, *Ocypode ryderi*.

Aerobic glycolysis occurs very rarely under physiological conditions, though it is a regular phenomenon in carcinoma cells because of their damaged mitochondria (Warburg et al. 1924). Its physiological significance is known in the special case of the so-called gas-gland in the swim bladder of teleost fishes without a ductus pneumaticus (physoclists; Berg and Steen 1968). The blood capillaries form loops orientated to the center of the bladder. Lactate built up in the gland liberates oxygen from hemoglobin in the blood by the counter-current principle (Bohr effect and Root effect functioning concomitantly), so that a very high partial pressure of free oxygen is formed in the apex of the loop. In the mammalian retina, capillary loops are known to occur from both networks of blood vessels on its inner and outer sides. Lactate could improve the oxygen supply to the retina substantially by this same mechanism. In insects, there is no need for a similar construction, for their blood transports only fuels, while oxygen is supplied as gas by the tracheoles which belong to each ommatidium.

Aerobic glycolysis occurring in the retina of Malacostraca must have an advantage for the retina different from that in mammals. The hemocyanin of *Ocypode* is known to have a Bohr effect, but addition of lactate to the hemolymph reduces the liberation of oxygen (Morris and Bridges 1985). Therefore, the biological significance of the aerobic glycolysis in both types of eyes cannot be understood in the same way. As an alternative hypothesis, it could be assumed that lactate may function as an easily accessible store of chemical energy which – in contrast to free glucose – does not inhibit the supply of glucose from the hemolymph by reducing the concentration difference.

References

Acker H, Holtermann G, Carlsson J, Nedermann T (1983) Methodological aspects of microelectrode measurements in cellular spheroids. Adv Exp Med Biol 159: 445–462

Autrum H, Tscharntke H (1962) Der Sauerstoffverbrauch der Insektenretina im Licht und im Dunkeln. Z Vergl Physiol 45: 695–710

Berg T, Steen JB (1968) The mechanism of oxygen concentration in the swim bladder of the eel. J Physiol (Lond) 195: 631–638

Cole WH (1941) A perfusing solution for the lobster *(Homarus)* heart and the effects of its constituent ions on the heart. J Gen Physiol 25: 1–6

Hamdorf K, Kaschef AH (1964) Der Sauerstoffverbrauch des Facettenauges von *Calliphora erythrocephala* in Abhängigkeit von der Temperatur und dem Ionenmilieu. Z Vergl Physiol 48: 251–265

Hamdorf K, Langer H (1966) Der Sauerstoffverbrauch des Facettenauges von *Calliphora erythrocephala* in Abhängigkeit von der Wellenlänge des Reizlichtes. Z Vergl Physiol 52: 386–400

Henning U, Langer H (1986) Untersuchungen zum Turnover der Photorezeptormembran im Auge der Krabbe *Ocypode ryderi*. Verh Dtsch Zool Ges 79: 213–214

Hoffmann C (1960) Belichtungspotentiale der Insekten und Sauerstoffdruck. Verh Dtsch Zool Ges 53: 220–225

Langer H (1962) Untersuchungen über die Größe des Stoffwechsels isolierter Augen von *Calliphora erythrocephala* Meigen. Biol Zentralbl 81: 691–720

Langer H, Lues I, Rivera ME (1976) Arginine phosphate in compound eyes. J Comp Physiol 107: 179–184

Lues I (1978) Untersuchungen zum Phosphat-Stoffwechsel der Facettenaugen decapoder Krebse. Dissertation, Fakultät für Biologie, Ruhr-Universität Bochum

Morris S, Bridges CR (1985) An investigation of haemocyanin oxygen affinity in the semi-terrestrial crab *Ocypode saratan* Forsk. J Exp Biol 117: 119–132

Rivera ME, Langer H (1983a) Enzyme pattern of energy releasing metabolism in eyes, optical ganglia and muscle of the crayfish *Astacus leptodactylus*. Mol Physiol 3: 313–329

Rivera ME, Langer H (1983b) Enzyme pattern of energy releasing metabolism in eyes and ganglia of the blowfly *Calliphora erythrocephala* and the crab *Ocypode ryderi*. Mol Physiol 4: 265–277

Sickel W (1972) Retinal metabolism in dark and light. In: Fuortes MGF (ed) Physiology of photoreceptor organs. Springer, Berlin Heidelberg New York, pp 667–727 (Handbook of sensory physiology, vol 7/2)

Tsacopoulos M, Poitry S, Borsellino A (1981) Diffusion and consumption of oxygen in the superfused retina of the drone *(Apis mellifera)* in darkness. J Gen Physiol 77: 601–628

Warburg O, Posener K, Negelein E (1924) Über den Stoffwechsel der Carcinomzelle. Biochem Z 152: 309–344

Subject Index

Acyl-CoA synthetase 97
Adenosine myocard 165
Adenylates heart 169
Adenylate cyclase 98
ADP liver cells 53
Aerobic glycolysis 41, 67, 205
Aerotaxis 7
Alonopine 27
Amine oxidase 54
Anaerobic glycolysis 15, 28, 38
    vertebrates, invertebrates 15
Angiogenesis factor 118
Anoxia
    different animals 15
    energy charge 17
    erythropoietin 108
    heart rate 17
    tolerance 20
    ventilation 16
Antenna system 4
Arachidonic acid 97, 104
Arenicola marina 27, 42
Aspartate 28
Autacoids endothelial cells 143
ATP bacteria 3
    carotid body 71
    central nervous system 17
    different organs 18
    endothelial cells 146
    heart 169
    kidney 94
    liver 53
    vertebrates, invertebrates 13
ATPase, calcium 73
Atta sexdens 26

Bacterial photosynthetic apparatus 3
Blood flow
    carotid body 66
    central nervous system 179
    heart 165
    kidney 103
    liver 50
Bohr effect 27
Burst-forming unit-erythroid 115
BW 755 C 136

Calcium
    carotid body 193
    endothelial cell 66, 147
    Influx type I cell 123
    liver 55
    mesangial cell 109
    smooth muscle 131
Carbon tetrachloride 58
Cardiomyocyte 165
Carotid body 65, 121, 193
    catecholamine 65, 121, 193
    cAMP 200
    in vitro 193
Cell-cycle progression 84
Chemosensory nerve response 196
Central nervous system 19, 179
Colon carcinoma 67
Colony-forming unit-erythroid 115
Compartimentation metabolism 95
Compound eye 205
Contractile functions 151
Cyclohexemide 89
Cyclooxygenase 104, 136
Cytochrome C 68
    oxidase 41, 51
    P 450 54, 58
Cytoskeleton 56

Decapod crustacea 205
Dithioerythrit 73
Dithiothreitol 144
Diving response 22
DNA gyrase 9
DNA replication 79
dopamine 125

Ehrlich ascites cells 79
Eicosanoid 108
Electromechanical coupling 138
Embden Meyerhof pathway 38
Endothelial cells
    coronary 166
    calcium 66
    hypoxia 66
    smooth muscle 136
Endothelium derived relaxant factor 132, 143

Energy charge, retina 208
Enzyme activities, retina 208
Erythropoietin
  different organs 116
  kidney 93, 103, 108
  macrophage 13
  stimulating factor 115
Escherichia coli 8

Fetal organs, erythropoietin 113
Fork progress rate 83
Fructose 2,6-biphosphate, different organs
  20

Ganglion cell sensory 179
Gene, erythropoietin 94
  fnr 8
  nif 9
Glioma cells 69, 123
Glutathione 73, 152
Glycogen aorta 7
Granulocyte-macrophage colony-forming cell
  114

Haloalkane 57
Halogenated hydrocarbons 58
Halothane 58
Helminths parasitic 31
Hemoglobin 146
Hemocyanin 205
Hippocampal slice 179
Hydrogen peroxide 53, 57
Hydroperoxides
  fatty acid 151
  low molecular 151
Hydroxyl radical 57
Hypercapnia
  central nervous system 179
  membrane characteristics 184
Hyperpolarisation, smooth muscle 131
Hyperoxia, membrane characteristics 187
Hypoxia
  action potential 180
  carotid body 65, 123, 193
  catecholamine synthesis 126
  cell death 54
  central nervous system 179
  endothelial cell 143
  environment 37
  erythropoietin 97
  liver cell 49
  membrane characteristics 133, 179
  myocard 165
  neuronal function 16
  smooth muscle 133
  tolerance 20
  vasodilatation 66, 131, 144

Iloprost 133
Indomethacin 135
Injury reoxygenation 56
Inositol phosphate 97
Insects 15
Interleukin-3 116
Invertebrates 13, 38
Ischemia
  brain 15
  kidney 105
  myocard 165

Lactate dehydrogenase 26
  isoenzymes 69
Levulinic acid 8
L-homocysteine 170
Light harvesting complexe 4
Lipid peroxidation 59
Lipooxygenase 132
Liver cell 49
LLC-PK 109
Locusta migratoria 25

Malacostraca 205
Malate dehydrogenase 27, 40
Membrane Potential, smooth muscle 131
Mesangial cell 108, 113
Monoamine oxidase 53
Motor cortex 179
Multicellular spheroid 67
Mytilus Edulis 42

Necrosis, centrilobular 54
^{31}P-NMR 29
Neuroblastoma cell 124

Ocypode ryderi 205
Opine 39
  dehydrogenase 27
Oxamic acid 70
Oxyconformers 37
Oxygen consumption
  carotid body 71
  liver cell 51
  PO_2-dependency 71, 200
  kidney 93
Oxygen diffusion
  intracellular 52
  limitation 46
Oxygen sensing
  bacteria 6
  carotid body 65, 121, 193
  central nervous system 179
  endothelial cell 143
  erythropoietin 93, 114
  invertebrates 13, 37
  kidney 93, 103

liver 49
  respiratory chain 69
  smooth muscle 132, 161
  vertebrates 18
Oxygen supply, kidney 93
Oxygen tension
  ATP synthesis 13, 37, 49, 70, 94
  bacterial photosynthesis 3
  bone marrow 115
  central nervous system 179
  critical 37
  DNA synthesis 79
  endothelial cells 66, 143
  erythropoietin 117
  gene expression 8
  heart 167
  hemopoiesis 114
  intracellular 50, 51
  kidney 106
  liver 50
  membrane differentiation 6
  multicellular spheroids 67
  smooth muscle 131, 151
Oxyregulators 37

Parasites 13, 38
Pasteur effect 19, 67
Peroxide dismutase 57
pH
  carotid body 71, 198
  central nervous system 182
  membrane characteristics 186
  multicellular spheroids 67
  retina 210
Phosphagen
  kinase 39
  utilization 38
Phosphoenolpyruvate 40
Phospholipids, liver 97
Phospholipase A 55
Phospholipase A_2 104
Phosphorylase, glycogen 154
Photosynthesis 4
PK-15 109
Pigment protein complexe 4
PIP kinase 96
Potassium
  carotid body 193
  central nervous system 181
  channel 96, 193
Procaryotic organism 3
Prostacyclin 132
Prostaglandine 97, 104
Prostanoids 103

Pyruvate kinase 40
Pyruvate reductase 39

Reaction center 5
Reoxygenation 56
Replicon cluster 83
Replicon initiation 80
Respiratory chain photometry 72
Retina 205
Rhodobacter capsulatus 5
Rhodobacter sphaeroides 6
mRNA erythropoietin 114

S-adenosylhomocysteine 170
Salmonella typhimurium 9
Scrobicularia plana 42
Shellfish 26
Sipunculus nudus, musculature 43
Skeletal muscle, white 38
Skin respiration 16
Smooth muscle 131, 151, 165
  sodium 131
S-phase program 79
Spinal cord 179
Stress, oxidative 151
Strombine 27, 39
Succinate 28, 38, 41
Succinate dehydrogenase 41
Superoxide anion radical 53

Tension, smooth muscle 131
Thiol, aorta 157
Thioredoxin 7
Tissue culture
  carotid body 121
  spinal cord 179
Tissue PO_2
  central nervous system 179
  heart 167
  kidney 106
  retina 210
Transduction mechanism 98
Transmitter secretion, carotid body 121, 193
Turtle 22
Type I cell 121, 193
Tyrosine hydroxylase 124

Urate oxidase 53
Urine production 107

Vertebrates 13

Xanthine oxidase 57